Quadratics

Explained like a friend would

—

For beginners :)

Hey, I need help with this quadratic stuff.

I just don't get it.

Alright.

Which bit?

All of it.

I don't even know what a quadratic is.

In class we're doing this thing called factoring.

But I have no idea what is going on.

Yeah.

There's a lot to learn with quadratics.

Different rules.

Different formulas.

You have to know when to use what.

And with factoring?

It's not straightforward.

It stumps a lot of people.

So, how do I learn this?

What's the easiest and the fastest way?

I feel like I've made 0 progress.

Look.

Like anything else.

Start with the basics.

Do some practice questions.

Build your understanding.

Once one thing makes sense.

Move to the next.

With quadratics.

Everything connects.

If you miss one part.

Or don't fully get it.

The other parts won't make sense at all.

You also need good algebra skills.

Fractions.

Exponents.

Surds.

All of that shows up in quadratics as well.

Yeah.

I got the *Algebra for Beginners* book.

Great book, by the way.

My algebra skills are actually pretty good now.

This is me shamelessly sneaking in a book recommendation here. ☺

Oh nice.

Because I was going to say to get that first.

Otherwise, this stuff can get hard to understand.

If your algebra's good.

Then let's get you good with quadratics too.

Stuff You're About to Get Real Good At

Expanding
Quadratics

Quadratics: When You Multiply a Linear By Another Linear

What's a quadratic?

A quadratic is what you get when you multiply two linear expressions together.

A *linear term* is basically a letter on its own.

Any letter. Doesn't matter which one.

It can also be multiplied or divided by any number.

Things like $x,\ 5y,\ 7z,\ \frac{a}{3},\ \frac{b}{2}$.

All of these are linear terms.

The important thing is that the letter has to have a power of 1.

A *linear expression* is either a linear term by itself.

Or a linear term with a number added or subtracted from it.

Now here's a small rule to keep in mind.

When you're doing something to an expression that has more than one term.

Adding, subtracting, multiplying, or dividing it.

That expression goes inside brackets.

So, with quadratics.

Being two linear expressions multiplied together.

They can be written in different ways.

It can be written with no brackets.

One bracket.

Or two brackets.

It just depends on whether each linear expression has one or two terms.

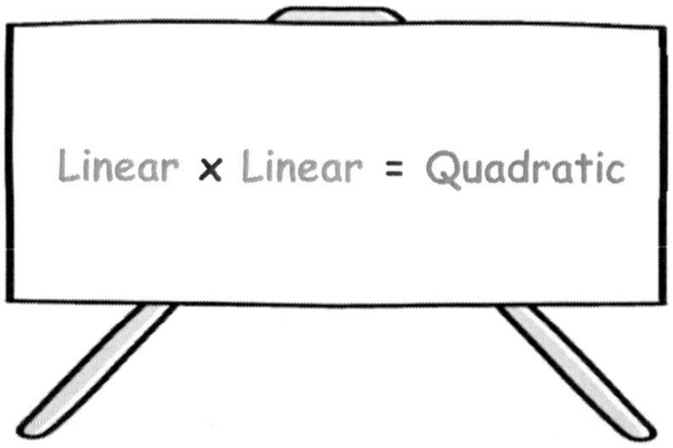

When written as the multiplication of two linear expressions.

This is called the *factored* form.

Which we'll get into later on.

You can also multiply out the two linear expressions.

And get rid of any brackets.

When you do that.

You'll end up with a quadratic that is said to be in **standard form**.

It looks like this.

The letter x is called the *variable*.

As mentioned before, it could be any letter.

$$ax^2 + bx + c$$

a, b and c represent numbers.

They can each be any number.

Positive or negative.

Except.

There's one crucial rule.

a can't be zero.

If a is zero.

It won't be a quadratic anymore.

Because the squared term gets multiplied by a.

And if a is zero, anything multiplied by zero is zero.

So, you'd end up with just $bx + c$.

Which is linear.

A quadratic has to have at least one term where the variable is squared.

It has a power of 2.

It doesn't need to have the middle linear term.

Or the number by itself.

That's why b and c can be zero.

But a can't.

In its simplest form, a quadratic is just one term.

A variable that's squared.

Quadratics can also have two or three terms.

As long as one of them is a squared variable term.

The other terms can either be a number or a linear term.

Nothing else, though.

You can't have variables with negative powers.

Fractional powers.

Or powers bigger than 2.

So, in short.

A quadratic is two linear expressions multiplied together.

It can be written with brackets.

Or it can be written in standard form, with no brackets.

Expanding Single Brackets: The Math Skill That Sticks Around (Like Forever)

Quadratics with single brackets come in two types.

First type.

A linear term outside.

A linear expression with two terms inside.

Second type.

A number outside.

A quadratic expression inside.

For both types.

Expanding the brackets is done in the same way.

You multiply what's outside by each term inside the brackets.

Starting with the first.

Then the second.

Then the third term, if there is one.

Here's the important bit.

The sign in front of each term belongs to that term.

Whether the term is inside or outside the brackets.

So, when multiplying, always include the sign.

Especially if it's negative.

That's where most of the mistakes happen.

If there's no sign, it just means it's positive.

Here's how to expand single brackets.

When you multiply the outside term by each term inside, do it in this order.

First, multiply the signs.

Then the numbers, if there are any.

Then, multiply the letter or letters.

If both terms have the same letter, the letter gets squared.

If only one term has a letter.

Multiply the numbers and write the letter with it.

If there's no letter, just multiply the numbers.

One thing to remember.

If there's no number in front of a letter, it means the number is 1.

We just don't write it.

Now, as you progress, you'll see two or more single brackets in one expression.

For these, expand each bracket separately.

Then combine like terms.

Like terms have the same letter and the same power.

For quadratics with the variable x.

Combine the x^2 terms together.

Then combine the x terms.

Then combine the numbers.

You can't combine x^2 with x.

They're not like terms.

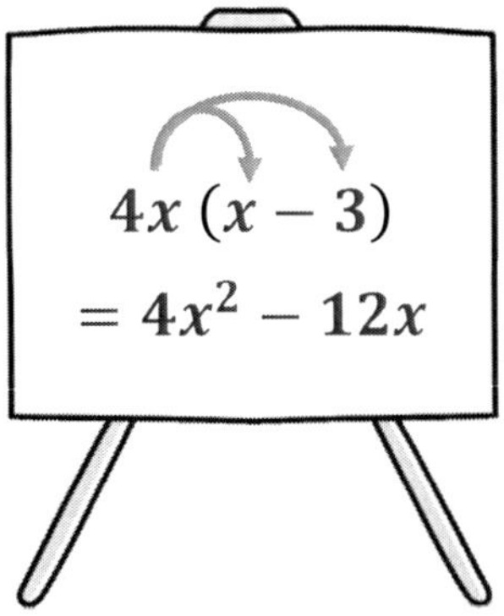

Worked Example 1

Expand the following.

a $x(x+2)$ b $y(3y-4)$ c $2z(z+5)$

d $3a(2a-7)$ e $4b(6-5b)$ f $-2c(-3c-4)$

Explanation		**Solution**
a	Write the expression.	$x(x+2)$
	Multiply the outside term by each inside term.	$= x^2 + 2x$
b	Write the expression.	$y(3y-4)$
	Multiply the outside term by each inside term.	$= 3y^2 - 4y$
c	Write the expression.	$2z(z+5)$
	Multiply the outside term by each inside term.	$= 2z^2 + 10z$
d	Write the expression.	$3a(2a-7)$
	Multiply the outside term by each inside term.	$= 6a^2 - 21a$
e	Write the expression.	$4b(6-5b)$
	Multiply the outside term by each inside term.	$= 24b - 20b^2$
f	Write the expression.	$-2c(-3c-4)$
	Multiply the outside term by each inside term.	$= 6c^2 + 8c$

Worked Example 2

Expand and simplify the following.

a $x(2x + 1) + x(x + 3)$

b $x(3x - 1) + x(2 - x)$

c $x(3 - x) - 2(x + 5)$

d $x(3x - 4) - 3(2 - 3x)$

Explanation		**Solution**
a	Write the expression.	$x(2x + 1) + x(x + 3)$
	Expand both brackets.	$= 2x^2 + x + x^2 + 3x$
	Combine like terms.	$= 3x^2 + 4x$
b	Write the expression.	$x(3x - 1) + x(2 - x)$
	Expand both brackets.	$= 3x^2 - x + 2x - x^2$
	Combine like terms.	$= 2x^2 + x$
c	Write the expression.	$x(3 - x) - 2(x + 5)$
	Expand both brackets.	$= 3x - x^2 - 2x - 10$
	Combine like terms.	$= -x^2 + x - 10$
d	Write the expression.	$x(3x - 4) - 3(2 - 3x)$
	Expand both brackets.	$= 3x^2 - 4x - 6 + 9x$
	Combine like terms.	$= 3x^2 + 5x - 6$

Worked Example 3

Expand the following.

a $2(x^2 + 4x + 5)$

b $3(4x^2 - 2x + 9)$

c $-3(2x^2 - 5x - 7)$

d $-5(8 + x - 2x^2)$

Explanation		Solution
a	Write the expression.	$2(x^2 + 4x + 5)$
	Multiply outside number by each term inside.	$= 2x^2 + 8x + 10$
b	Write the expression.	$3(4x^2 - 2x + 9)$
	Multiply outside number by each term inside.	$= 12x^2 - 6x + 27$
c	Write the expression.	$-3(2x^2 - 5x - 7)$
	Multiply outside number by each term inside.	$= -6x^2 + 15x + 21$
d	Write the expression.	$-5(8 + x - 2x^2)$
	Multiply outside number by each term inside.	$= -40 - 5x + 10x^2$
		$= 10x^2 - 5x - 40$

When multiplying or dividing, same signs give a positive answer, different signs give a negative answer.

Worked Example 4

Expand and simplify the following.

a $x(x+4)+5(x^2+1)$ b $x(3-2x)+4(x^2-2)$

	Explanation	**Solution**
a	Write the expression.	$x(x+4)+5(x^2+1)$
	Expand both brackets.	$=x^2+4x+5x^2+5$
	Combine like terms.	$=6x^2+4x+5$
b	Write the expression.	$x(3-2x)+4(x^2-2)$
	Expand both brackets.	$=3x-2x^2+4x^2-8$
	Combine like terms.	$=3x+2x^2-8$
		$=2x^2+3x-8$

After expanding a more than one bracket in one expression, look for like terms. Then combine and simplify.

Worked Example 5

Expand and simplify the following.

a $3(x^2 + 4) + x(4 - x)$

b $2(1 - x^2) - 3x(5 + 2x)$

Explanation	**Solution**
a Write the expression.	$3(x^2 + 4) + x(4 - x)$
Expand both brackets.	$= 3x^2 + 12 + 4x - x^2$
Combine like terms.	$= 2x^2 + 4x + 12$
b Write the expression.	$2(1 - x^2) - 3x(5 + 2x)$
Expand both brackets.	$= 2 - 2x^2 - 15x - 6x^2$
Combine like terms.	$= 2 - 8x^2 - 15x$
	$= -8x^2 - 15x + 2$

Quadratics are usually written with x^2 term first, then x term and the number last. (Or whatever the variable is)

Worked Example 6

Expand and simplify the following.

a $2(3x+1)+5(x^2-2x+1)$

b $-6(x^2+3x-1)-4x(7-2x)$

c $x(5-2x)-7(-x^2+3x-1)$

d $-6(2x-5-x^2)-2x(4-x)+x$

Explanation	**Solution**
a Write the expression.	$2(3x+1)+5(x^2-2x+1)$
Expand both brackets.	$=6x+2+5x^2-10x+5$
Combine like terms.	$=-4x+7+5x^2$
	$=5x^2-4x+7$
b Write the expression.	$-6(x^2+3x-1)-4x(7-2x)$
Expand both brackets.	$=-6x^2-18x+6-28x+8x^2$
Combine like terms.	$=2x^2-46x+6$
c Write the expression.	$x(5-2x)-7(-x^2+3x-1)$
Expand both brackets.	$=5x-2x^2+7x^2-21x+7$
Combine like terms.	$=5x^2-16x+7$
d Write the expression.	$-6(2x-5-x^2)-2x(4-x)+x$
Expand both brackets.	$=-12x+30+6x^2-8x+2x^2+x$
Combine like terms.	$=8x^2-19x+30$

Worked Example 7

Find the value of k for each of the following.

a $4(3x - x^2) + 5x(x - 6) = x^2 - kx$

b $-5(-x^2 + x - 4) - 2x(3 - x) + 6x^2 = 13x^2 + kx + 20$

Explanation	Solution
a Write the expression.	$4(3x - x^2) + 5x(x - 6)$
Expand both brackets.	$= 12x - 4x^2 + 5x^2 - 30x$
Combine like terms.	$= x^2 - 18x$
Let the two expressions equal.	$x^2 - 18x = x^2 - kx$
Match the coefficients.	$\therefore\ k = 18$
b Write the expression.	$-5(-x^2 + x - 4) - 2x(3 - x) + 6x^2$
Expand both brackets.	$= 5x^2 - 5x + 20 - 6x + 2x^2 + 6x^2$
Combine like terms.	$= 13x^2 - 11x + 20$
Let the two expressions equal.	$13x^2 - 11x + 20 = 13x^2 + kx + 20$
Match the coefficients.	$\therefore\ k = -11$

Your Turn To Try

1 **Expand $x(2x + 3)$. What's the result?**

A $x^2 + 3x$ B $2x^2 - 3x$

C $2x^2 + 3$ D $2x^2 + 3x$

2 **What is the result of expanding $3x(4x - 7)$?**

A $12x^2 - 21$ B $12x^2 - 21x$

C $12x - 21$ D $12x^2 + 21x$

3 **What is the expanded form of $-4a(8 - 5a)$?**

A $20a^2 - 32a$ B $-20a^2 + 32a$

C $20a^2 + 32a$ D $-20a^2 - 32a$

4 **Expand $2x(x - 2) - x(x + 4)$. What do you get?**

A $3x^2$ B x^2

C $x^2 - 8x$ D $x^2 + 8x$

5 **Find the result of expanding: $x(5 - x) - 3(x^2 - 1)$?**

A $4x^2 + 5x + 3$ B $-4x^2 + 5x + 3$

C $4x^2 - 5x + 3$ D $-4x^2 - 5x + 3$

Question	1	2	3	4	5
Answer	D	B	A	C	B

Time To Practice

1 Expand the following.

a $x(x+5)$	b $y(2y+1)$	c $a(3a-1)$
d $2x(1-x)$	e $3y(2-y)$	f $-2a(2+3a)$
g $-3x(6x-5)$	h $-5y(1-2y)$	i $-4a(3-4a)$

2 Expand and simplify the following.

a $x(x+3)+x(x+4)$	b $x(x-5)+x(x-7)$
c $x(3x-1)+x(4-x)$	d $x(4x-3)-2x(x-7)$
e $x(5x+2)+2x(6-x)$	f $3x(2x+4)-x(5-2x)$
g $4x(2x-1)+2x(1-3x)$	h $2x(2-3x)-3x(2x-7)$

3 Expand and simplify the following.

a $3(2x^2-5x+4)$	b $-2(x^2-3x-7)$
c $4(x-2-3x^2)$	d $-5(9-8x+6x^2)$

4 Expand and simplify the following.

a $3x(5-2x)-2(x^2-3)$	b $4x(1+2x)-3(x^2+3)$
c $x(7-x)-5(x^2-1)$	d $-2x(1-3x)-4(x^2-6)$
e $3(2x-7)+7(x^2-x+4)$	f $-2(4x+3)-(-5x^2-3x+4)$
g $6(1+3x-2x^2)-2x(3x+1)$	h $-5(-x^2+x+3)+2x(1-x)$
i $-7(6x^2+x-8)+2x(1+3x)$	j $-3(-4x^2-x-2)-3(-x-5)$

Challenge Yourself

1 The expanded version of $3x(2-x)-4(1-x^2)$ is kx^2+6x-4.
What is the value of k?

2 Simplifying $-4y^2+2y(3y-5)+6(y^2-y)$, the result is $8y^2-ky$.
Find k?

3 The expanded form of $2(3p-p^2)+p(4p-1)-5(p+2)$ is kp^2-10.
Determine the value of k?

4 The expansion of $3(-x^2+x+2)+2x(x-4)$ is $-x^2-5x+k$.
What is the value of k?

5 The simplified form of $5(2n-n^2)+3n(n-7)-4(-n^2+3n-2)$ is $2n^2-kn+8$. What is the value of k?

6 Simplifying $6t^2-3t(t+2)+4(t^2-t)+3t$ is kt^2-kt.
What is the value of k?

Expanding Double Brackets: Every Term Gets a Turn To Multiply

This is when two brackets are multiplied together.
Inside each, there are two terms.

Expanding these works just like single brackets.
Except you're multiplying twice.

Here's how.

Multiply the first term in the first bracket by each term in the second.
Then multiply the second term in the first bracket by each term in the second.
Once everything's multiplied.
Simplify by combining like terms.

Remember.
The sign in front of each term goes with that term.
So, when multiplying, always include the sign.
Specially the second term inside each bracket.

If there's more than one pair of double brackets.
Keep any numbers or signs outside the brackets as they are.
Just expand each pair separately.
Write the result inside its own single bracket.
From there, multiply what's outside by this new single bracket.
Then simplify by combining like terms.

Worked Example 1

Expand and simplify the following.

a $(x+2)(x+3)$ b $(x-9)(x+4)$ c $(x-5)(x-7)$

Explanation		**Solution**
a	Write the expression.	$(x+2)(x+3)$
	Multiply each term in the first bracket by each term in the second.	$= x^2 + 3x + 2x + 6$
	Combine like terms.	$= x^2 + 5x + 6$
b	Write the expression.	$(x-9)(x+4)$
	Multiply each term in the first bracket by each term in the second.	$= x^2 + 4x - 9x - 36$
	Combine like terms.	$= x^2 - 5x - 36$
c	Write the expression.	$(x-5)(x-7)$
	Multiply each term in the first bracket by each term in the second.	$= x^2 - 7x - 5x + 35$
	Combine like terms.	$= x^2 - 12x + 35$

The two **numbers** inside the brackets, with their signs, multiply to give the **number** by itself in the standard form.

Worked Example 2

Expand and simplify the following.

a $(4 - x)(7 - x)$ b $(2x + 6)(x + 5)$ c $(x - 9)(9x - 1)$

Explanation	Solution
a Write the expression.	$(4 - x)(7 - x)$
Multiply each term in the first bracket by each term in the second.	$= 28 - 4x - 7x + x^2$
Combine like terms.	$= x^2 - 11x + 35$
b Write the expression.	$(2x + 6)(x + 5)$
Multiply each term in the first bracket by each term in the second.	$= 2x^2 + 10x + 6x + 30$
Combine like terms.	$= 2x^2 + 16x + 30$
c Write the expression.	$(x - 9)(9x - 1)$
Multiply each term in the first bracket by each term in the second.	$= 9x^2 - x - 81x + 9$
Combine like terms.	$= 9x^2 - 82x + 9$

Worked Example 3

Expand and simplify the following.

a $(3x+5)(6x-7)$ b $2(x+6)(3x+5)$ c $-5(7x-4)(2-x)$

Explanation		Solution
a	Write the expression.	$(3x+5)(6x-7)$
	Multiply each term in the first bracket by each term in the second.	$=18x^2-21x+30x-35$
	Combine like terms.	$=18x^2+9x-35$
b	Write the expression.	$2(x+6)(3x+5)$
	Expand the double brackets.	$=2(3x^2+5x+18x+30)$
	Combine like terms inside.	$=2(3x^2+23x+30)$
	Multiply outside number by each term inside.	$=6x^2+46x+60$
c	Write the expression.	$-5(7x-4)(2-x)$
	Expand the double brackets.	$=-5(14x-7x^2-8+4x)$
	Combine like terms inside.	$=-5(18x-7x^2-8)$
	Multiply outside number by each term inside.	$=-90x+35x^2+40$
		$=35x^2-90x+40$

Worked Example 4

Expand and simplify the following.

a $(2x-5)(5x-3)-8x$

b $5x-(3x+1)(2x-1)$

Explanation	Solution
a Write the expression.	$(2x-5)(5x-3)-8x$
Expand the double brackets and simplify.	$=\left(10x^2-6x-25x+15\right)-8x$ $=\left(10x^2-31x+15\right)-8x$
Combine like terms.	$=10x^2-31x+15-8x$ $=10x^2-39x+15$
b Write the expression.	$5x-(3x+1)(2x-1)$
Expand the double brackets and simplify.	$=5x-\left(6x^2-3x+2x-1\right)$ $=5x-\left(6x^2-x-1\right)$
Multiply the negative outside to each term inside.	$=5x-6x^2+x+1$
Combine like terms.	$=-6x^2+6x+1$

When there's no number in front of a single or double bracket, it's the same as having a 1 in front of it. We just don't write the 1.

Worked Example 5

Expand and simplify $(d-7)(d+1)-(5d-3)(d+4)$

Explanation	**Solution**
Write the expression.	$(d-7)(d+1)-(5d-3)(d+4)$
Expand the first double brackets.	$(d-7)(d+1)$ $= d^2+d-7d-7$ $= d^2-6d-7$
Expand the second double brackets.	$(5d-3)(d+4)$ $= 5d^2+20d-3d-12$ $= 5d^2+17d-12$
Subtract the two expanded quadratics. The second one goes inside brackets.	$= d^2-6d-7-(5d^2+17d-12)$
Multiply the negative outside to each term inside.	$= d^2-6d-7-5d^2-17d+12$
Combine like terms.	$= -4d^2-23d+5$

Always use brackets when adding or subtracting anything with more than one term to avoid mistakes with negatives.

Worked Example 6

Expand and simplify $(-4x-5)(3x-2)-(-x+6)(5x+3)$

Explanation	**Solution**
Write the expression.	$(-4x-5)(3x-2)-(-x+6)(5x+3)$
Expand the first double brackets.	$(-4x-5)(3x-2)$ $=-12x^2+8x-15x+10$ $=-12x^2-7x+10$
Expand the second double brackets.	$(-x+6)(5x+3)$ $=-5x^2-3x+30x+18$ $=-5x^2+27x+18$
Subtract the two expanded quadratics. The second one goes inside brackets.	$=-12x^2-7x+10-(-5x^2+27x+18)$
Multiply the negative outside to each term inside.	$=-12x^2-7x+10+5x^2-27x-18$
Combine like terms.	$=-7x^2-34x-8$

A **variable term** is a term that contains a letter.

Worked Example 7

Expand and simplify $-(p-8)(3p+1)-2(p+3)(2p-7)$

Explanation	**Solution**
Write the expression.	$-(p-8)(3p+1)-2(p+3)(2p-7)$
Expand the first double brackets.	$(p-8)(3p+1)$ $=3p^2+p-24p-8$ $=3p^2-23p-8$
Expand the second double brackets.	$(p+3)(2p-7)$ $=2p^2-7p+6p-21$ $=2p^2-p-21$
Subtract the two expanded quadratics. Write each result inside brackets.	$=-(3p^2-23p-8)-2(2p^2-p-21)$
Multiply the negative numbers outside to each term inside.	$=-3p^2+23p+8-4p^2+2p+42$
Combine like terms.	$=-7p^2+25p+50$

A number by itself, with no variable is called the **constant.**

Worked Example 8

Find the value of k for each of the following, where k is a constant

a $(2x - 5)(x + k) = 2x^2 + x - 15$ b $(kx + 2)(3x + 7) = 6x^2 + 17x - 14$

c $(1 - 8x)(2x + 9) = -16x^2 + kx + 9$

Explanation	**Solution**
a Write the equation.	$(2x - 5)(x + k) = 2x^2 + x - 15$
Multiply the constants inside each bracket. This equals the constant in standard form.	$-5 \times k = -15$
Solve for k.	$k = \frac{-15}{-5}$ $k = 3$
b Write the equation.	$(kx + 2)(3x + 7) = 6x^2 + 17x - 14$
Multiply variable terms inside each bracket. This equals the squared term.	$3kx^2 = 6x^2$
Solve for k.	$3k = 6$ $k = 2$
c Write the equation.	$(1 - 8x)(2x + 9)$
Expand the brackets.	$= 2x + 9 - 16x^2 - 72x$ $= -16x^2 - 70x + 9$
Let the two expressions equal.	$-16x^2 - 70x + 9 = -16x^2 + kx + 9$
Match the coefficients.	$\therefore k = -70$

Your Turn To Try

1 **When expanded the expression $(x + 3)(x - 4)$ simplifes to?**

A $x^2 - x + 12$

B $x^2 - x - 12$

C $x^2 + x - 12$

D $x^2 - 12$

2 **What is the result of expanding $(x - 2)(x - 3)$?**

A $x^2 - 6x + 5$

B $x^2 - 3x - 2$

C $x^2 - 5x + 6$

D $x^2 + 5x + 6$

3 **Which is the correct expanded result of $(x - 6)(2x + 3)$?**

A $2x^2 - 15x + 18$

B $2x^2 - 9x + 18$

C $2x^2 - 9x - 18$

D $2x^2 - 12x - 18$

4 **What will you get after expanding and simplifying $(4x - 1)(2x + 5)$?**

A $8x^2 + 14x - 5$

B $8x^2 + 14x + 5$

C $8x^2 + 18x + 5$

D $8x^2 + 18x - 5$

5 **Find the result of expanding $2(x - 2)(x - 3)$?**

A $2x^2 - 10x + 12$

B $2x^2 + 2x - 6$

C $2x^2 - 2x - 6$

D $2x^2 + 10x + 12$

Question	1	2	3	4	5
Answer	B	C	C	D	A

Time To Practice

1 Expand and simplify the following.

a $(x+3)(x+4)$ b $(x+2)(x+8)$ c $(x+8)(x+3)$

d $(p-2)(p+3)$ e $(p-5)(p+4)$ f $(p+3)(p-7)$

g $(k-4)(k-6)$ h $(k-5)(k-8)$ i $(k-1)(k-9)$

2 Expand and simplify the following.

a $(2x+1)(x+5)$ b $(3x-1)(x+1)$ c $(2x-3)(x+2)$

d $(4p+5)(3p+2)$ e $(3p+2)(3p-5)$ f $(5p+3)(4p-2)$

g $(3+2k)(7-k)$ h $(5-2k)(3+4k)$ i $(11-3k)(10-7k)$

3 Expand and simplify the following.

a $2(x+1)(x+3)$ b $3(x+2)(x-7)$

c $2(p-8)(p+1)$ d $4(p-9)(p-2)$

e $-5(m+3)(m+2)$ f $-4(m-1)(m+6)$

g $-2(x-3)(x-8)$ h $3(2x+1)(2x+3)$

i $6(3y-4)(y+2)$ j $-2(z+4)(3z-7)$

4 Expand and simplify the following.

a $(x+2)(x-1)-2x$ b $5x+(x+3)(x-4)$

c $3x-(2x-5)(x+2)$ d $2x-(3x-1)(2x-3)$

e $(3-7x)(2x-5)-4x$ f $-7x-(3x-1)(2x-9)$

Time To Practice

5 Expand and simplify the following.

a $(x+1)(x+3)+(x+2)(x+4)$

b $(a+3)(a-1)+(a-2)(a-4)$

c $(2a-7)(a+4)-(a+6)(a-3)$

d $(y+3)(y-5)-(y+6)(2y+5)$

e $(4b+8)(b+5)-(3b-5)(b-7)$

f $(3c-5)(2c+7)-(3c+5)(c-4)$

g $(4x-1)(3x+5)-(7x-9)(2x-1)$

h $(5p-3)(p-7)+(5p+3)(p+7)$

i $(9-2y)(3-4y)-(3y+2)(1-5y)$

j $(6-5r)(3r-2)-(-7r-8)(1+4r)$

k $(4w+9)(-2w-3)+(3w-1)(1-3w)$

l $(-3q-4)(3-5q)-(5q+7)(-q-3)$

m $-4(2a-3)(a+2)-(3a-4)(a-1)$

n $3(5-2s)(s+2)-2(s+1)(s-2)$

o $-7(2y+5)(y+1)-2(1-4y)(y+3)$

Challenge Yourself

1 What is the value of k?

a $(x+4)(x+k) = x^2 + 11x + 28$

b $(x+7)(x-k) = x^2 + 4x - 21$

c $(x-k)(x-8) = x^2 - 10x + 16$

2 Expand and simplify the following.

a $\left(x - 3\sqrt{2}\right)\left(2x - 2\sqrt{2}\right)$

b $\frac{1}{3}(1-3x)(2x-5)$

c $\frac{(x+1)(2x-3)}{3} - \frac{(2x-1)(x+3)}{5}$

Expanding Perfect Squares: The Identical Twin Brackets

Expanding double brackets.

Like we just did.

Works for all types of double brackets.

But there are two special cases where you don't need to expand this way.

There's a shortcut formula instead.

The first special case is when both brackets are *exactly* the same.

Same terms.

Same signs.

They're called *perfect squares*.

Depending on the sign inside the brackets.

Whether it's a plus or a minus.

You use one of these formulas to expand the brackets.

$$(a+b)(a+b) = a^2 + 2ab + b^2$$

$$(a-b)(a-b) = a^2 - 2ab + b^2$$

a and b can be anything.

One's a variable term, the other's a number.

Looking at the expanded versions.

- There are three terms.
- The first one comes from squaring the first term inside the brackets.
- The third one comes from squaring the second term inside the brackets.
- Both of these terms are *always* positive.
- The middle term is 2 multiplied by the first and second terms inside the brackets.
- It's sign matches whatever's inside the brackets. Plus or minus.

Remember these six points.

Because you'll need them later when factoring perfect squares.

You could expand these like normal double brackets.

Multiplying each term in the first bracket by everything in the second.

But using these formulas?

Way quicker.

You get the answer in one step.

Just keep in mind, this only works when both brackets are exactly the same.

Here's another way you'll see perfect squares written.

Same formula.

The brackets are just squared.

$$(a + b)^2 = a^2 + 2ab + b^2$$

$$(a - b)^2 = a^2 - 2ab + b^2$$

In the first couple of examples we'll go through every step.

So you can see how it works.

Then we'll expand in one step.

That's where you want to be.

To use the formula.

1) Square whatever a is. That becomes the first term when expanded.
2) Multiply a and b and then multiply what you get by 2. That's the middle term.
3) Add or subtract this middle term.
4) Square whatever b is. That becomes the third term when expanded.

Worked Example 1

Expand the following expressions.

a $(x + 8)(x + 8)$ b $(x - 5)(x - 5)$ c $(3 - 2y)(3 - 2y)$

Explanation		Solution
a	Write the expression.	$(x + 8)(x + 8)$
	Identify a and b and square each term.	$a = x \text{ and } b = 8$ $a^2 = x^2 \text{ and } b^2 = 64$
	Work out $2ab$.	$2ab = 2(x)(8) = 16x$
	Use the perfect square formula.	$= x^2 + 16x + 64$
b	Write the expression.	$(x - 5)(x - 5)$
	Identify a and b and square each term.	$a = x \text{ and } b = 5$ $a^2 = x^2 \text{ and } b^2 = 25$
	Work out $2ab$.	$2ab = 2(x)(5) = 10x$
	Use the perfect square formula.	$= x^2 - 10x + 25$
c	Write the expression.	$(3 - 2y)(3 - 2y)$
	Identify a and b and square each term.	$a = 3 \text{ and } b = 2y$ $a^2 = 9 \text{ and } b^2 = 4y^2$
	Work out $2ab$.	$2ab = 2(3)(2y) = 12y$
	Use the perfect square formula.	$= 9 - 12y + 4y^2$

Worked Example 2

Expand the following expressions.

a $(m+2)^2$ b $(n-1)^2$ c $(10-p)^2$

Explanation	Solution
a Write the expression.	$(m+2)^2$
Identify a and b and square each term.	$a = m \text{ and } b = 2$ $a^2 = x^2 \text{ and } b^2 = 4$
Work out $2ab$.	$2ab = 2(m)(2) = 4m$
Use the perfect square formula.	$= m^2 + 4m + 4$
b Write the expression.	$(n-1)^2$
Identify a and b and square each term.	$a = n \text{ and } b = 1$ $a^2 = n^2 \text{ and } b^2 = 1$
Work out $2ab$.	$2ab = 2(n)(1) = 2n$
Use the perfect square formula.	$= n^2 - 2n + 1$
c Write the expression.	$(10-p)^2$
Identify a and b and square each term.	$a = 10 \text{ and } b = p$ $a^2 = 100 \text{ and } b^2 = p^2$
Work out $2ab$.	$2ab = 2(10)(p) = 20p$
Use the perfect square formula.	$= 100 - 20p + p^2$

Worked Example 3

Expand and simplify the following expressions.

a $(2x+1)(2x+1)$ b $(4y-1)^2$ c $(2-3b)^2$

	Explanation	Solution
a	Write the expression.	$(2x+1)(2x+1)$
	Use the perfect square formula.	$=(2x)^2+2(2x)(1)+(1)^2$
	Simplify.	$=4x^2+4x+1$
b	Write the expression.	$(4y-1)^2$
	Use the perfect square formula.	$=(4y)^2-2(4y)(1)+(1)^2$
	Simplify.	$=16y^2-8y+1$
c	Write the expression.	$(2-3b)^2$
	Use the perfect square formula.	$=(2)^2-2(2)(3b)+(3b)^2$
	Simplify.	$=4-12b+9b^2$

When expanding $(\boldsymbol{a}-\boldsymbol{b})^2$ using the formula, don't include the negative with the $\boldsymbol{b}$ term. The formula already takes care of that minus for you.

Worked Example 4

Expand and simplify $(2x-3)^2-(2x+1)(3x+9)$

Explanation	Solution
a Write the expression.	$(2x-3)^2-(2x+1)(3x+9)$
Expand first brackets using the perfect square formula.	$(2x-3)^2=4x^2-12x+9$
Expand second brackets.	$(2x+1)(3x+9)$ $=6x^2+18x+3x+9$ $=6x^2+21x+9$
Subtract the two expanded quadratics.	$=4x^2-12x+9-\left(6x^2+21x+9\right)$
Multiply the negative numbers outside to each term inside.	$=4x^2-12x+9-6x^2-21x-9$
Combine like terms.	$=-2x^2-33x$

Forgetting the middle term when expanding perfect squares.
$(a+b)^2\neq a^2+b^2$ or $(a-b)^2\neq a^2-b^2$.

Worked Example 5

The simplified form of $(x+1)^2 - kx - 4 = x^2 - x - 3$.

Find k where k is a constant.

Explanation	Solution
Write the equation.	$(x+1)^2 - kx - 4 = x^2 - x - 3$
Expand the brackets on the left.	$= x^2 + 2x + 1 - kx - 4$
Collect like terms.	$= x^2 + (2-k)x - 3$
Let the two expressions equal.	$x^2 + (2-k)x - 3 = x^2 - x - 3$
Match the coefficients.	$2 - k = -1$
Solve for k.	$2 + 1 = k$
	$\therefore\ k = 3$

Matching the coefficients is the same thing as **equating the coefficients**.
It's just the simpler way to say it.

Expanding Difference of Two Squares: The Not-So-Identical Twin Brackets

The second special case is called the *Difference of Two Squares* (DOTS).

It's when there are two brackets with the exact same terms inside.

Except the signs are different.

One bracket has a minus.

The other has a plus.

$$(a + b)(a - b) = a^2 - b^2$$

Pay attention to these little details.

Because again, when you're factoring, you'll need to know what to look for.

In the expanded version.

- There are only two terms
- Both are perfect squares
- The first comes from squaring the first term inside the brackets
- The second comes from squaring the second term
- There's always minus in between them

Notice there's no middle term.

That's because when you expand the brackets, the middle terms cancel each other out.

How to use this formula.

1) Square whatever a is. That becomes the first term when expanded.
2) Square whatever b is. That becomes the second term.
3) Put a minus sign in between.

Worked Example 6

Expand and simplify the following.

a $(x-2)(x+2)$

b $(a+3)(a-3)$

c $(2x-7)(2x+7)$

d $(4a+5)(4a-5)$

Explanation	Solution
a Write the expression.	$(x-2)(x+2)$
Use the DOTS formula.	$= x^2-(2)^2$
Simplify.	$= x^2-4$
b Write the expression.	$(a+3)(a-3)$
Use the DOTS formula.	$= a^2-(3)^2$
Simplify.	$= a^2-9$
c Write the expression.	$(2x-7)(2x+7)$
Use the DOTS formula.	$= (2x)^2-(7)^2$
Simplify.	$= 4x^2-49$
d Write the expression.	$(4a+5)(4a-5)$
Use the DOTS formula.	$= (4a)^2-(5)^2$
Simplify.	$= 16a^2-25$

Worked Example 7

Expand and simplify the following.

a $5(2x-3)(2x+3)$

b $(3-4y)(3+4y)+2$

Explanation	Solution
a Write the expression.	$5(2x-3)(2x+3)$
Use the DOTS formula to expand the brackets.	$=5[(2x)^2-(3)^2]$
	$=5(4x^2-9)$
Multiply the number outside to each term inside.	$=20x^2-45$
b Write the expression.	$(3-4y)(3+4y)+2$
Use the DOTS formula to expand the brackets.	$=(3)^2-(4y)^2+2$
Simplify and combine like terms.	$=9-16y^2+2$
	$=11-16y^2$

Your Turn To Try

1 **If you expand $(4m+1)^2$ what will the middle term be?**

A $8m$
B $4m$
C $16m$
D $2m$

2 **Expand and simplify $(x-3)^2$. What do you get?**

A x^2-6x-9
B x^2-3x+9
C x^2-9
D x^2-6x+9

3 **Which is the correct expansion of $(3-2p)(3-2p)$?**

A $9-6p-p^2$
B $4p^2-12p+9$
C $4p^2-6p+9$
D $2p^2-6p-9$

4 **What is the expanded form of $(5x-4)(5x+4)$?**

A $25x^2-16$
B $25x^2-40p+16$
C $25x^2+16$
D $25x^2-40p-16$

5 **Which of can be expanded using the difference of two squares formula?**

A $(2x+5)(2x-1)$
B $(1-9y)(1+9y)$
C $(4-x)(4-x)$
D $(6x-7)(6x-7)$

Question	1	2	3	4	5
Answer	A	D	B	A	B

Time To Practice

1 Expand the following.

a $(x-4)^2$ b $(5-y)^2$ c $(a+2)^2$

d $5(x+3)^2$ e $-2(y-4)^2$ f $-2(a+6)^2$

g $(x+7)^2-5x$ h $2(y+3)^2-10$ i $-(2a+5)^2-10a$

2 Expand the following.

a $(x+6)(x+6)$ b $(y+3)(y+3)$ c $(7-a)(7-a)$

d $3(x+2)(x+2)$ e $2(y-5)(y-5)$ f $-6(a-2)(a-2)$

3 Expand and simplify the following.

a $(2x-3)(2x-3)-9$ b $14-(2x+5)(2x+5)$

c $5-(3x-1)(3x-1)$ d $2-(3-11x)(3-11x)$

e $90x^2-(2-9x)(2-9x)$ f $(3+4x)(3+4x)-2x^2$

4 Expand the following.

a $(x+7)(x-7)$ b $(y+9)(y-9)$ c $(5-a)(5+a)$

d $(3x-1)(3x+1)$ e $2(y-5)(y+5)$ f $-6(a-2)(a+2)$

5 Expand and simplify the following.

a $3(2x-5)(2x+5)$ b $2(4y-3)(4y+3)+6$

c $-(7p-3)(7p+3)+12$ d $5-(2-9q)(2+9q)$

e $3x^2-(6-5x)(6+5x)$ f $(4z-1)(4z+1)-6z^2$

Challenge Yourself

1 Expand and simplify the following.

a $\left(x-\frac{1}{2}\right)^2$

b $\left(\frac{3}{2}x+\frac{1}{3}\right)^2$

c $\left(5x-\frac{2}{3}\right)\left(5x+\frac{2}{3}\right)$

d $2\left(\frac{3}{4}x+\frac{1}{2}\right)\left(\frac{3}{4}x-\frac{1}{2}\right)$

2 Expand and simplify the following.

a $4(x-2)^2-(x-1)^2$

b $\frac{1}{3}(6-7x)(6+7x)-\frac{2}{3}x^2$

c $\left(\sqrt{2}-3x\right)\left(\sqrt{2}+3x\right)$

d $\left(\sqrt{2}x-1\right)\left(\sqrt{2}x+1\right)$

3 The simplified form of $(x-2)^2-5x+k=x^2-9x+13$.
Find k where k is a constant.

4 The simplified form of $(5x+1)^2-2(x-k)=25x^2+8x+9$.
Find k where k is a constant.

5 The simplified form of $(x+1)(x-1)+k(3x+1)=x^2-18x-7$.
Find k where k is a constant.

1. Answers

Single Brackets

Time to Practice

1 a $x^2 + 5x$ b $2y^2 + y$ c $3a^2 - a$

d $2x - 2x^2$ e $6y - 3y^2$ f $-4a - 6a^2$

g $-18x^2 + 15x$ h $-5y + 10y^2$ i $-12a + 16a^2$

2 a $2x^2 + 7x$ b $2x^2 - 12x$ c $2x^2 + 3x$

d $2x^2 + 11x$ e $3x^2 + 14x$ f $8x^2 + 7x$

g $2x^2 - 2x$ h $-12x^2 + 25x$

3 a $6x^2 - 15x + 12$ b $-2x^2 + 6x + 14$ c $-12x^2 + 4x - 8$

d $-30x^2 + 40x - 45$

4 a $-8x^2 + 15x + 6$ b $5x^2 + 4x - 9$ c $-6x^2 + 7x + 5$

d $2x^2 - 2x + 24$ e $7x^2 - x + 7$ f $5x^2 - 5x - 10$

g $-18x^2 + 16x + 6$ h $3x^2 - 3x - 15$ i $-36x^2 - 5x + 56$

j $12x^2 + 6x + 21$

Challenge Yourself

1 $k = 1$ 2 $k = 16$ 3 $k = 2$

4 $k = 6$ 5 $k = 23$ 6 $k = 7$

Double Brackets

Time to Practice

1	a	$x^2 + 7x + 12$	b	$x^2 + 10x + 16$	c	$x^2 + 11x + 24$
	d	$p^2 + p - 6$	e	$p^2 - p - 20$	f	$p^2 - 4p - 21$
	g	$k^2 - 10k + 24$	h	$k^2 - 13k + 40$	i	$k^2 - 10k + 9$
2	a	$2x^2 + 11x + 5$	b	$3x^2 + 2x - 1$	c	$2x^2 + x - 6$
	d	$12p^2 + 23p + 10$	e	$9p^2 - 9p - 10$	f	$20p^2 + 2p - 6$
	g	$-2k^2 + 11k + 21$	h	$-8k^2 + 14k + 15$	i	$21k^2 - 107k + 110$
3	a	$2x^2 + 8x + 6$	b	$3x^2 - 15x - 42$	c	$2p^2 - 14p - 16$
	d	$4p^2 - 44p + 72$	e	$-5m^2 - 25m - 30$	f	$-4m^2 - 20m + 24$
	g	$-2x^2 + 22x - 48$	h	$12x^2 + 24x + 9$	i	$18y^2 + 12y - 48$
	j	$-6z^2 - 10z + 56$				
4	a	$x^2 - x - 2$	b	$x^2 + 4x - 12$	c	$-2x^2 + 4x + 10$
	d	$-6x^2 + 13x - 3$	e	$-14x^2 + 37x - 15$	f	$-6x^2 + 22x - 9$
5	a	$2x^2 + 10x + 11$	b	$2a^2 - 4a + 5$	c	$a^2 - 2a - 10$
	d	$-y^2 - 19y - 45$	e	$b^2 + 54b + 5$	f	$3c^2 + 18c - 15$
	g	$-2x^2 + 42x - 14$	h	$10p^2 + 42$	i	$23y^2 - 35y + 25$
	j	$13r^2 + 67r - 4$	k	$-17w^2 - 24w - 28$	l	$20q^2 + 33q + 9$
	m	$-11a^2 + 3a + 20$	n	$-8s^2 + 5s + 34$	o	$-6y^2 - 27y - 41$

Challenge Yourself

1	a	7	b	3	c	2
2	a	$2x^2 + 8x\sqrt{2} + 12$	b	$-2x^2 + \frac{17x}{3} - \frac{5}{3}$	c	$\frac{4x^2 - 20x - 6}{15}$

Perfect Squares and Difference of Two Squares

Time to Practice

1 a $x^2 - 8x + 16$ b $25 - 10y + y^2$ c $a^2 + 4a + 4$

d $5x^2 + 30x + 45$ e $-2y^2 + 16y - 32$ f $-2a^2 - 24a - 72$

g $x^2 + 9x + 49$ h $2y^2 + 12y + 8$ i $-4a^2 - 30a - 25$

2 a $x^2 + 12x + 36$ b $y^2 + 6y + 9$ c $49 - 14a + a^2$

d $3x^2 + 12x + 12$ e $2y^2 - 20y + 50$ f $-6a^2 + 24a - 24$

3 a $4x^2 - 12x$ b $1x^2 - 20x - 11$ c $-9x^2 + 6x + 4$

d $-121x^2 + 66x - 7$ e $9x^2 + 36x - 4$ f $14x^2 + 24x + 9$

4 a $x^2 - 49$ b $y^2 - 81$ c $25 - a^2$

d $9x^2 - 1$ e $2y^2 - 50$ f $-6a^2 + 24$

5 a $12x^2 - 75$ b $32y^2 - 12$ c $-49p^2 + 21$

d $1 + 81q^2$ e $28x^2 - 36$ f $10z^2 - 1$

Challenge Yourself

1 a $x^2 - x + \frac{1}{4}$ b $\frac{9}{4}x^2 + x + \frac{1}{9}$ c $25x^2 - \frac{4}{9}$

d $\frac{9}{8}x^2 - \frac{1}{2}$

2 a $3x^2 - 14x + 15$ b $12 - 17x^2$ c $2 - 9x^2$

d $2x^2 - 1$

3 $k = 9$ 4 $k = 4$ 5 $k = -6$

Factoring
Quadratics

Factoring Quadratics: Where the Brackets Make a Comeback

You've just learned how to expand brackets.

Single.

Double.

Perfect squares.

And the difference of two squares.

Factoring is the opposite of expanding.

You go from the expanded form back to the bracket form.

Either to a single bracket.

Or to double brackets.

This bracket form is called the *factored form*.

To get to this form, there are 5 main techniques.

1. Greatest Common Factor
2. Perfect Squares
3. Difference of Two Squares
4. AC Method
5. Completing the Square

That's the order to check when factoring quadratics.

There's no formula.

Where you just plug in numbers and factor.

You need to know which technique to use.

And when to use it.

If you follow these 5 techniques in order.

It removes the guesswork.

And works every time.

Here's what you do.

Start with the first technique.

If it doesn't work, move to the next.

Keep going until you find the one that works.

What?

Go through every single technique?

It's much quicker than it sounds.

You can tell straight away if the first three will work or not.

Just by looking at the quadratic.

Most of the time you skip them completely.

The AC method?

You'll use this one the most.

As for the last one, completing the square.

It's not the fastest or easiest way to factor.

You only use this technique if the AC Method doesn't work.

Or if the question specifically asks you to.

If it doesn't ask.

Don't use it.

Greatest Common Factor: Where Factoring Begins (And Sometimes Ends)

The first thing to do whenever factoring quadratics.

Check if it has a Greatest Common Factor (GCF).

If it does.

Factor it out.

If it doesn't.

Move to the next technique.

Sometimes factoring out the GCF is all you need to do.

Other times, it simplifies the quadratic.

Making it easier to factor further using another technique.

Which is why you always check for a GCF first.

To find the GCF, look at the numbers in each term.

The biggest number that goes into all of them.

That's the GCF.

If the quadratic has two terms.

Where one has a squared letter and the other also has the letter.

Then that letter's part of the GCF too.

If the second term is a number.

Or the quadratic has three terms, then the letter won't be part of the GCF.

The GCF will just be a number.

Once you've found the GCF.

Write it outside a set of brackets.

Inside the brackets, divide each term by that GCF.

That's what it means to "factor it out".

To know if you've factored correctly.

You can expand the brackets.

If you get back the original quadratic.

You did it right.

If you don't, something's wrong.

Go back and double check.

You don't have to do this.

Expand and check.

It's just a way to make sure you did it right.

Maybe do it until you get the hang of it.

In this section, the questions can only be factored by finding the GCF.

But when we get into the other techniques, you'll need to factor further.

After factoring out the GCF.

In harder questions.

The GCF can be an expression.

It's not always a number or a letter.

So, keep an eye out for anything that's in every term.

Usually written in brackets.

One term will have that bracket squared.

The other will have the same bracket by itself, with a number in front.

Alright, here are the steps.

1) Find the GCF
2) Write it outside a pair of brackets
3) Divide each term by the GCF and write what's left inside the brackets

Worked Example 1

Factor the following expressions.

a $4x^2 + 8$

b $6x^2 + 15$

c $18 + 12x^2$

d $5p^2 - 35$

e $8q^2 - 6$

f $63 - 14r^2$

Explanation	Solution
a Write the expression.	$4x^2 + 8$
Factor out the GFC 4.	$= 4(x^2 + 2)$
b Write the expression.	$6x^2 + 15$
Factor out the GFC 3.	$= 3(2x^2 + 5)$
c Write the expression.	$18 + 12x^2$
Factor out the GFC 6.	$= 6(3 + 2x^2)$
d Write the expression.	$5p^2 - 35$
Factor out the GFC 5.	$= 5(p^2 - 7)$
e Write the expression.	$8q^2 - 6$
Factor out the GFC 2.	$= 2(4q^2 - 3)$
f Write the expression.	$63 - 14r^2$
Factor out the GFC 7.	$= 7(9 - 2r^2)$

Worked Example 2

Factor the following expressions.

a $x^2 + x$ b $2x^2 + 10x$ c $12x + 8x^2$

d $10p^2 - 100p$ e $18q^2 - 30q$ f $33r - 22r^2$

Explanation		Solution
a	Write the expression.	$x^2 + x$
	Factor out the GFC x.	$= x(x + 1)$
b	Write the expression.	$2x^2 + 10x$
	Factor out the GFC $2x$.	$= 2x(x + 5)$
c	Write the expression.	$12x + 8x^2$
	Factor out the GFC $4x$.	$= 4x(3 + 2x)$
d	Write the expression.	$10p^2 - 100p$
	Factor out the GFC $10p$.	$= 10p(p - 10)$
e	Write the expression.	$18q^2 - 30q$
	Factor out the GFC $6q$.	$= 6q(3q - 5)$
f	Write the expression.	$33r - 22r^2$
	Factor out the GFC $11r$.	$= 11r(3 - 2r)$

Worked Example 3

Factorise the following.

a $\mathbf{3x^2 + 12x + 21}$

b $\mathbf{2x^2 - 4x + 4}$

c $\mathbf{18x^2 - 24x + 10}$

d $\mathbf{-16x^2 + 16x - 8}$

Explanation	Solution
a Write the expression.	$3x^2 + 12x + 21$
Factor out the GFC 3.	$= 3(x^2 + 4x + 7)$
b Write the expression.	$2x^2 - 4x + 4$
Factor out the GFC 2.	$= 2(x^2 - 2x + 2)$
c Write the expression.	$18x^2 - 24x + 10$
Factor out the GFC 2.	$= 2(9x^2 - 12x + 5)$
d Write the expression.	$-16x^2 + 16x - 8$
Factor out the GFC -8.	$= -8(2x^2 - 2x + 1)$

If you see a negative in front of the squared term, take -1 out as a factor.

Worked Example 4

Factorise the following.

a $3(x+2) - x(x+2)$

b $(a-6) - a(a-6)$

c $-x(x+4) - 3(x+4)$

d $-12(2y+7) - y(2y+7)$

Explanation	Solution
a Write the expression.	$3(x+2) - x(x+2)$
Factor out the GFC $(x+2)$.	$= (x+2)(3-x)$
b Write the expression.	$(a-6) - a(a-6)$
Factor out the GFC $(a-6)$.	$= (a-6)(1-a)$
c Write the expression.	$-x(x+4) - 3(x+4)$
Factor out the GFC $(x+4)$.	$= (x+4)(-x-3)$
Factor out negative from second bracket.	$= -(x+4)(x+3)$
d Write the expression.	$-12(2y+7) - y(2y+7)$
Factor out the GFC $(2y+7)$.	$= (2y+7)(-12-y)$
Factor out negative from second bracket.	$= -(2y+7)(y+12)$

Worked Example 5

Factorise the following by taking out the greatest common factor.

a $(p+1)^2+3(p+1)$

b $(2n-1)^2-4(2n-1)$

c $-(x+9)^2-3(x+9)$

Explanation	**Solution**
a Write the expression.	$(p+1)^2+3(p+1)$
Factor out the GFC $(p+1)$.	$=(p+1)[(p+1)+3]$
Combine like terms and simplify second bracket.	$=(p+1)(p+4)$
b Write the expression.	$(2n-1)^2-4(2n-1)$
Factor out the GFC $(2n-1)$.	$=(2n-1)[(2n-1)-4]$
Combine like terms and simplify second bracket.	$=(2n-1)(2n-5)$
c Write the expression.	$-(x+9)^2-3(x+9)$
Factor out the GFC $-(x+9)$.	$=-(x+9)[(x+9)+3]$
Combine like terms and simplify second bracket.	$=-(x+9)(x+12)$

When factoring out a negative, forgetting to change the sign of all the terms inside the brackets.

Your Turn To Try

1 **What is the GCF of $6x^2 + 9$?**

A 2 B 9

C 3 D 6

2 **What is the factored form of $15x^2 + 30$?**

A $5(3x^2 + 6)$ B $15(x^2 + 2)$

C $30(x^2 + 1)$ D $3(5x^2 + 10)$

3 **In the expression $16x^2 - 4x$, what is the GCF?**

A $2x$ B $8x$

C $4x$ D 4

4 **Factoring $8p - 24p^2$. What do you get?**

A $8(p - 3p^2)$ B $4p(2 - 6p)$

C $4(2p - 6p^2)$ D $8p(1 - 3p)$

5 **What is factored form of $18x^2 - 12x + 3$?**

A $6(3x^2 - 2x + 3)$ B $3(6x^2 + 4x + 1)$

C $3(6x^2 - 4x - 1)$ D $3(6x^2 - 4x + 1)$

Question	1	2	3	4	5
Answer	C	B	C	D	D

Time To Practice

1 Factorise by taking out the common factors.

a $2x^2 - 4$ b $3a^2 + 9$ c $-12b^2 - 3$

d $15x^2 + 25$ e $27 - 9a^2$ f $-32b^2 - 48$

g $75x^2 - 100$ h $35a^2 - 49$ i $81b^2 - 27$

2 Factor the following quadratics.

a $6x^2 + 8x$ b $4a^2 - 2a$ c $14b^2 - 21b$

d $12x - 12x^2$ e $9a^2 - 15a$ f $18b - 4b^2$

g $2x^2 - 32x$ h $12a^2 - 27a$ i $-5b - 2b^2$

3 Factorise the following quadratics by taking out the greatest common factor.

a $2x^2 + 4x + 6$ b $3a^2 - 6a - 12$ c $12b^2 + 24b + 36$

d $5x^2 - 10x + 15$ e $6a^2 - 18a + 24$ f $8b^2 + 8b - 32$

g $-4x^2 - 8x - 12$ h $-15a^2 + 30a - 45$ i $-9b^2 + 27b - 36$

4 Factor the following quadratics.

a $9(x + 1) + x(x + 1)$ b $x(x + 5) - 2(x + 5)$

c $-4(y - 7) + y(y - 7)$ d $2y(y - 1) - 3(y - 1)$

e $p(3p + 1) - 8(3p + 1)$ f $-5(2p + 3) - 7p(2p + 3)$

g $2x(4x - 9) - 3(4x - 9)$ h $-(3x - 5) - 4x(3x - 5)$

Time To Practice

5 Factorise the following quadratics by taking out the greatest common factor.

a $(x+2)^2+3(x+2)$

b $(y-5)^2-4(y-5)$

c $-2(x+4)^2+7(x+4)$

d $2(y-1)^2-9(y-1)$

e $(4x-1)^2-2(4x-1)$

f $(3y+1)^2+(3y+1)$

g $(3-2x)^2+2(3-2x)$

h $-4(2y+5)^2+3(2y+5)$

Challenge Yourself

1 Factorise the following quadratics.

a a^2b-ab^2

b $2x^2yz-4xy$

c $-12m^2n-12mn^2$

d $6xyz^2-3yz^2$

e $-2p^2q+4pq-6p^2$

f $5a^2b-10ab^2-15ab$

g $3(x-1)(x+2)+2(x-1)$

h $5(y-2)-10(2y-1)(y-2)$

Perfect Square Trinomials: When Three Terms Become One (Squared)

When factoring a three-term quadratic.

First thing.

Check if there's a GCF.

If there's one.

Factor it out.

You'll have a simpler quadratic inside brackets.

Check if that's a perfect square.

If there's no GCF to begin with.

Check the quadratic as it is.

Is it a perfect square?

If it is.

Use the perfect square formula to factor it.

The one used for expanding two identical brackets.

Only this time, it's used backwards.

$$a^2 + 2ab + b^2 = (a + b)^2$$

$$a^2 - 2ab + b^2 = (a - b)^2$$

If the quadratic is not a perfect square.

Either the original.

Or the simpler one after factoring out the GCF.

Try the AC Method.

How do you know if it's a perfect square?

Remember the 6 points from expanding identical brackets?

- There are three terms.
- The first one comes from squaring the first term inside the brackets.
- The third one comes from squaring the second term inside the brackets.
- Both of these terms are always positive.
- The middle term is 2 multiplied by the first and second terms inside the brackets.
- Its sign matches whatever's inside the brackets. Plus or minus.

For a perfect square.

All of these must be true.

Even if one isn't.

It won't be a perfect square.

Here's the fastest way to check.

Is the quadratic three terms?

Yes.

Are the first and third terms positive?

Yes.

Are the first and third terms perfect squares?

Yes.

Is the middle term 2 times the square roots of the first and third terms?

Yes.

If ALL of these are yes.

It's a perfect square.

If any of these is a no.

It's not a perfect square.

Applying the formula backwards.

Here are the steps to factor a perfect square.

1) Write a set of brackets and square it
2) The square root of the first term goes inside as the first term
3) The square root of the third term goes in as the second term
4) If the middle term's positive, write a plus between them. If negative, write a minus

For the first couple of examples.

We'll do the full check.

So you know what to look for.

Towards the end, we'll factor them in one step.

Worked Example 1

Factorise the following.

a $x^2 - 18x + 81$ b $x^2 + 12x + 36$

Explanation		Solution
a	Write the expression.	$x^2 - 18x + 81$
	Three terms?	Yes
	Last term is positive?	Yes
	First and last terms are perfect squares?	Yes. (x^2 and 81)
	Middle term is $2ab$?	$2 \times x \times 9 = 18x$
	Factor by using the perfect square formula.	$= (x - 9)^2$
b	Write the expression.	$x^2 + 12x + 36$
	Three terms?	Yes
	Last term is positive?	Yes
	First and last terms are perfect squares?	Yes. (x^2 and 36)
	Middle term is $2ab$?	$2 \times x \times 6 = 12x$
	Factor by using the perfect square formula.	$= (x + 6)^2$

You might hear these called perfect square trinomials.
'Tri' means three, for the three terms.

Worked Example 2

Factorise the following.

a $4x^2 + 20x + 25$

b $9x^2 - 24x + 16$

Explanation		Solution
a	Write the expression.	$4x^2 + 20x + 25$
	Three terms?	Yes
	Last term is positive?	Yes
	First and last terms are perfect squares?	Yes. ($4x^2$ and 25)
	Middle term is $2ab$?	$2 \times 2x \times 5 = 20x$
	Factor by using the perfect square formula.	$= (2x + 5)^2$
b	Write the expression.	$9x^2 - 24x + 16$
	Three terms?	Yes
	Last term is positive?	Yes
	First and last terms are perfect squares?	Yes. ($9x^2$ and 16)
	Middle term is $2ab$?	$2 \times 3x \times 4 = 24x$
	Factor by using the perfect square formula.	$= (3x - 4)^2$

If the quadratic term is negative, always factor out the negative first.

Worked Example 3

Factorise the following.

a $x^2 - 20x + 100$ b $4x^2 - 28x + 49$ c $36 - 60a + 25a^2$

Explanation	Solution
a Write the expression.	$x^2 - 20x + 100$
Rewrite the first and third terms as something squared.	$= x^2 - 20x + 10^2$
Factor by using the perfect square formula.	$= (x - 10)^2$
b Write the expression.	$4x^2 - 28x + 49$
Rewrite the first and third terms as something squared.	$= (2x)^2 - 28x + (7)^2$
Factor by applying the formula.	$= (2x - 7)^2$
c Write the expression.	$36 - 60a + 25a^2$
Rewrite the first and third terms as something squared.	$= 6^2 - 60a + (5a)^2$
Factor by using the perfect square formula.	$= (6 - 5a)^2$

If you rewrite the first and third terms as something squared, it makes it easier to identify 'a' and 'b' in the formula. But you don't need to do this step.

Worked Example 4

Factorise the following.

a $\mathbf{3x^2 - 30x + 75}$ b $\mathbf{18x^2 + 12x + 2}$ c $\mathbf{-50x^2 + 40x - 8}$

Explanation	**Solution**
a Write the expression.	$3x^2 - 30x + 75$
Factor out the GCF.	$= 3(x^2 - 10x + 25)$
Rewrite the first and third terms as something squared.	$= 3(x^2 - 10x + 5^2)$
Factor by using the perfect square formula.	$= 3(x - 5)^2$
b Write the expression.	$18x^2 + 12x + 2$
Factor out the GCF.	$= 2(9x^2 + 6x + 1)$
Rewrite the first and third terms as something squared.	$= 2((3x)^2 + 6x + 1^2)$
Factor by using the perfect square formula.	$= 2(3x + 1)^2$
c Write the expression.	$-50x^2 + 40x - 8$
Factor out the GCF.	$= -2(25x^2 - 20x + 4)$
Rewrite the first and third terms as something squared.	$= -2((5x)^2 - 20x + 2^2)$
Factor by using the perfect square formula.	$= -2(5x - 2)^2$

Perfect Squares 2: The Squared Part Wants To Be Anything But x

Some of the harder looking quadratics.

The squared term isn't a letter.

It's an expression being squared.

And the middle term also has that expression.

For these, if they have the pattern of a perfect square.

Factor them the same way.

But instead of a letter as the first term inside the brackets.

It'll be the expression instead.

Once factored.

Simplify by combining like terms.

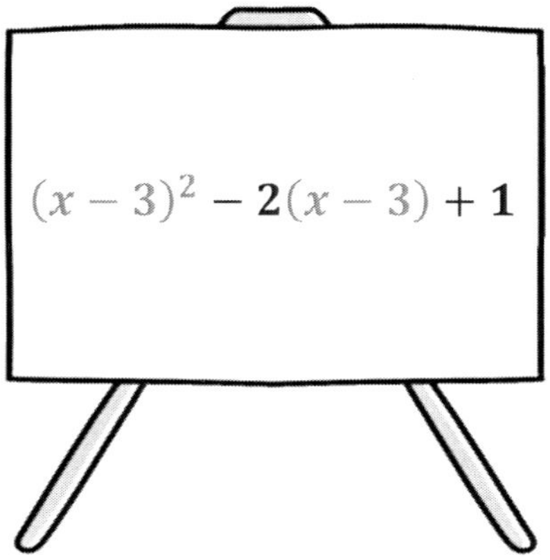

Another way to factor these is by *substitution*.

Here's what you do.

Look at what's being squared.

Pick a letter and let it equal that expression.

Just not the same one already in the question.

Most people use the letter u.

But it can be anything.

Then 'substitute' that letter in place of the expression in the quadratic.

This makes it easier to work with.

Because after the substitution it looks like a normal quadratic.

From there, factor it.

Once it's factored.

Swap the letter back with the original expression.

Tidy it up by collecting like terms.

Sometimes you'll see an expression and think.

This isn't even a quadratic.

The highest power isn't 2.

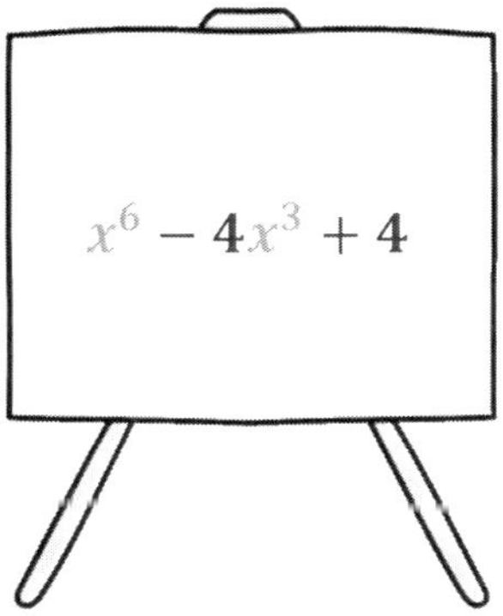

Yes.

True.

Technically not a quadratic.

But if it's set up like a quadratic.

It can be treated and factored like one.

What does that mean?

The variable's highest power is twice the power of the middle term.

For these, you can also use substitution.

Let the middle variable, with its power, equal a new letter.

Say the letter u.

This turns the expression into a regular quadratic in terms of u.

Now you can factor it like normal.

Once you've factored it, swap u back to what it equaled.

If it turns out to be a perfect square.

The same six points mentioned before will be true.

Factor it by applying the formula backwards.

Then swap u back to what it equaled.

Worked Example 5

Factorise $(x-1)^2 + 4(x-1) + 4$

Explanation	Solution
Write the expression.	$(x-1)^2 + 4(x-1) + 4$
Rewrite the first and third terms as something squared.	$= (x-1)^2 + 4(x-1) + 2^2$
Factor by using the perfect square formula.	$= \left((x-1) + 2\right)^2$
Simplify.	$= (x-1+2)^2$
	$= (x+1)^2$
	or
Write the expression.	$(x-1)^2 + 4(x-1) + 4$
Use the substitution method.	Let $u = x - 1$
Rewrite quadratic in terms of u.	$= u^2 + 4u + 4$
Rewrite the first and third terms as something squared.	$= u^2 + 4u + 2^2$
Factor by applying the formula.	$= (u+2)^2$
Replace u with original expression.	$= (x-1+2)^2$
Simplify.	$= (x+1)^2$

Worked Example 6

Factorise $(2x-3)^2 - 8(2x-3) + 16$

Explanation	Solution
Write the expression.	$(2x-3)^2 - 8(2x-3) + 16$
Rewrite the first and third terms as something squared.	$= (2x-3)^2 - 8(2x-3) + 4^2$
Factor by using the perfect square formula.	$= [(2x-3) - 4]^2$
Simplify.	$= (2x-3-4)^2$
	$= (2x-7)^2$
	or
Write the expression.	$(2x-3)^2 - 8(2x-3) + 16$
Use the substitution method.	Let $u = 2x - 3$
Rewrite quadratic in terms of u.	$= u^2 - 8u + 16$
Rewrite the first and third terms as something squared.	$= u^2 - 8u + 4^2$
Factor by using the perfect square formula.	$= (u-4)^2$
Replace u with original expression.	$= (2x-3-4)^2$
Simplify.	$= (2x-7)^2$

Worked Example 7

Factorise the following.

a $x^4 - 6x^2 + 9$ b $p^6 + 4p^3 + 4$ c $4z^8 + 24z^4 + 25$

Explanation		Solution
a	Write the expression.	$x^4 - 6x^2 + 9$
	Rewrite the first and third terms as something squared.	$= (x^2)^2 - 6x^2 + 3^2$
	Factor by using the perfect square formula.	$= (x^2 - 3)^2$
b	Write the expression.	$p^6 + 4p^3 + 4$
	Rewrite the first and third terms as something squared.	$= (p^3)^2 + 4p^3 + 2^2$
	Factor by using the perfect square formula.	$= (p^3 + 2)^2$
c	Write the expression.	$4z^8 + 20z^4 + 25$
	Rewrite the first and third terms as something squared.	$= (2z^4)^2 + 20z^4 + 5^2$
	Factor by using the perfect square formula.	$= (2z^4 + 5)^2$

Your Turn To Try

1 **Which of the following is the factored form of $x^2 + 6x + 9$?**

A $(x - 3)(x - 3)$ B $(x + 3)(x + 3)$

C $(x + 1)(x + 9)$ D $(x + 2)(x + 3)$

2 **What is the factored form of $x^2 - 16x + 64$?**

A $(x - 8)(x - 8)$ B $(x + 2)(x + 8)$

C $(x - 2)(x - 8)$ D $(x + 8)(x + 8)$

3 **Identify the factored form of $-x^2 - 12x - 36$.**

A $(x + 6)^2$ B $(x - 6)^2$

C $-(x + 6)^2$ D $-(x - 6)^2$

4 **Select the factored form of $4x^2 - 20x + 25$.**

A $(2x + 5)^2$ B $(2x - 5)^2$

C $(5 + 2x)^2$ D $(5 - 2x)^2$

5 **Factor $16 - 48x + 36x^2$. What do you get?**

A $4(3 - 2x)^2$ B $4(3 + 2x)^2$

C $4(2 + 3x)^2$ D $4(2 - 3x)^2$

Question	1	2	3	4	5
Answer	B	A	C	B	D

Time To Practice

1 Factor each of the following quadratics.

a $x^2 + 14x + 49$ b $a^2 - 6a + 9$ c $b^2 + 2b + 1$

d $x^2 - 4x + 4$ e $-a^2 - 18a - 81$ f $25 - 10b + b^2$

g $-x^2 - 16x - 64$ h $-a^2 + 22a - 121$ i $-144 + 24b - b^2$

2 Factor the following quadratics.

a $9x^2 + 12x + 4$ b $16a^2 - 40a + 25$ c $-25b^2 + 30b - 9$

d $-49x^2 - 28x - 4$ e $36a^2 + 84a + 49$ f $-81b^2 + 90b - 25$

g $100x^2 - 60x + 9$ h $-4a^2 - 20a - 25$ i $1 + 22b + 121b^2$

3 Factor the following quadratics.

a $3x^2 + 30x + 75$ b $3a^2 - 24a + 48$ c $2b^2 - 16b + 32$

d $4x^2 + 24x + 36$ e $-9a^2 + 54a - 81$ f $100 - 100b + 25b^2$

g $-16x^2 - 32x - 16$ h $10a^2 - 60a + 90$ i $-50 + 140b - 98b^2$

Challenge Yourself

1 Factorise the following expressions.

a $(x + 5)^2 - 12(x + 5) + 36$ b $(a + b)^2 + 12(a + b) + 36$

c $x^4 + 10x^2 + 25$ d $4a^8 - 28a^4 + 49$

e $-16x^6 + 56x^3 - 49$ f $9x^4 - 12x^2y^2 + 4y^4$

Difference of Two Squares: When Minus Signs Are a Plus (For Factoring)

Factoring a Difference of Two Squares (DOTS) is the easiest.

In terms of knowing when to use it.

And doing the factoring too.

If the quadratic has two terms.

A squared term minus a number.

Or a squared term minus another squared term.

Use this technique.

If the quadratic has anything else.

You can't use this technique.

Try another one.

To factor the difference of two squares.

Use the same formula we used when expanding.

Only backwards.

$$a^2 - b^2 = (a + b)(a - b)$$

Here are the steps to factor.

1) Write two brackets side by side.
2) First term in each bracket is the square root of the first term of the expression.
3) Second term in each bracket is the square root of the second term of the expression.
4) Write a plus in one bracket and a minus in the other.

Doesn't matter which bracket has the plus or the minus sign.

It works either way.

With some of the trickier DOTS.

The squared term will be a bracket that's squared.

Factoring these works the same way.

Just take the square root of the whole bracket.

Then follow the same 4 factoring steps above.

You'll just need to tidy up by combining like terms.

Here are a few tips.

When you take the square root of a squared bracket.

The square root of that is just whatever's inside the brackets.

And this will often have more than one term.

Because of that, when you write it inside the two factored brackets.

Put it inside its own brackets too.

So you'll end up with brackets inside brackets.

From there, simplify again.

Lastly, if it makes it easier to see what's being squared.

You can rewrite each term as something squared.

Before you start factoring.

Oh, and don't forget to factor out the greatest common factor first.

If there is one.

Worked Example 1

Factorise the following.

a $a^2 - 4$

b $b^2 - 36$

c $c^2 - 25$

d $x^2 - y^2$

e $1 - y^2$

f $9 - z^2$

Explanation		Solution
a	Write the expression.	$a^2 - 4$
	Factor using the DOTS formula.	$= (a - 2)(a + 2)$
b	Write the expression.	$b^2 - 36$
	Factor using the DOTS formula.	$= (b - 6)(b + 6)$
c	Write the expression.	$c^2 - 25$
	Factor using the DOTS formula.	$= (c - 5)(c + 5)$
d	Write the expression.	$x^2 - y^2$
	Factor using the DOTS formula.	$= (x - y)(x + y)$
e	Write the expression.	$1 - y^2$
	Factor using the DOTS formula.	$= (1 - y)(1 + y)$
f	Write the expression.	$9 - z^2$
	Factor using the DOTS formula.	$= (3 - z)(3 + z)$

Worked Example 2

Factorise the following.

a $\mathbf{4a^2 - 1}$ b $\mathbf{16b^2 - 49}$ c $\mathbf{100 - 9y^2}$

Explanation		**Solution**
a	Write the expression.	$4a^2 - 1$
	Rewrite each term as something squared.	$= (2a)^2 - 1^2$
	Factor using the DOTS formula.	$= (2a - 1)(2a + 1)$
b	Write the expression.	$16b^2 - 49$
	Rewrite each term as something squared.	$= (4b)^2 - 7^2$
	Factor using the DOTS formula.	$= (4b - 7)(4b + 7)$
c	Write the expression.	$100 - 9y^2$
	Rewrite each term as something squared.	$= 10^2 - (3y)^2$
	Factor using the DOTS formula.	$= (10 - 3y)(10 + 3y)$

Using the DOTS technique to factor when two terms that are added. ⚠

Worked Example 3

Factorise the following.

a $2a^2 - 72$ b $18b^2 - 50$ c $12c^2 - 27$

Explanation		**Solution**
a	Write the expression.	$2a^2 - 72$
	Take out the GCF.	$= 2(a^2 - 36)$
	Rewrite each term as something squared.	$= 2(a^2 - 6^2)$
	Factor using the DOTS formula.	$= 2(a - 6)(a + 6)$
b	Write the expression.	$18b^2 - 50$
	Take out the GCF.	$= 2(9b^2 - 25)$
	Rewrite each term as something squared.	$= 2[(3b)^2 - 5^2]$
	Factor using the DOTS formula.	$= 2(3b - 5)(3b + 5)$
c	Write the expression.	$12c^2 - 27$
	Take out the GCF.	$= 3(4c^2 - 9)$
	Rewrite each term as something squared.	$= 3[(2c)^2 - 3^2]$
	Factor using the DOTS formula.	$= 3(2c - 3)(2c + 3)$

Worked Example 4

Factorise the following.

a $(x+3)^2-9$ b $(3x-2)^2-81$ c $64-(2-3x)^2$

Explanation	Solution
a Write the expression.	$(x+3)^2-9$
Rewrite each term as something squared.	$=(x+3)^2-3^2$
Factor using the DOTS formula.	$=(x+3-3)(x+3+3)$
Simplify.	$=x(x+6)$
b Write the expression.	$(3x-2)^2-81$
Rewrite each term as something squared.	$=(3x-2)^2-9^2$
Factor using the DOTS formula.	$=(3x-2-9)(3x-2+9)$
Simplify.	$=(3x-11)(3x+7)$
c Write the expression.	$64-(2x-3)^2$
Rewrite each term as something squared.	$=8^2-(2x-3)^2$
Factor using the DOTS formula.	$=\big(8-(2x-3)\big)\big(8+(2x-3)\big)$
Remove inner brackets.	$=(8-2x+3)(8+2x-3)$
Simplify.	$=(11-2x)(2x+5)$
Optional. Take out a negative from the first bracket.	$=-(2x-11)(2x+5)$

Worked Example 5

Factorise the following.

a $-3(x-4)^2+27$

b $4(3x-1)^2-(x+2)^2$

Explanation	**Solution**
a Write the expression.	$-3(x-4)^2+27$
Factor out the GCF.	$=-3[(x-4)^2-9]$
Rewrite each term as something squared.	$=-3[(x-4)^2-3^2]$
Factor using the DOTS formula.	$=-3(x-4-3)(x-4+3)$
Simplify.	$=-3(x-7)(x-1)$
b Write the expression.	$4(3x-1)^2-(x+2)^2$
Rewrite each term as something squared.	$=[2(3x-1)]^2-(x+2)^2$
Factor using the DOTS formula.	$=\big(2(3x-1)-(x+2)\big)\big(2(3x-1)+(x+2)\big)$
Expand inner brackets.	$=(6x-2-x-2)(6x-2+x+2)$
Simplify.	$=(5x-4)(7x)$
	$=7x(5x-4)$

If the first term is negative and the second is positive, you can swap the order of the terms or factor out a negative 1 first. Both ways work.

Worked Example 6

Factorise the following.

a $p^2 - \frac{9}{49}$ b $\frac{x^2}{25} - \frac{1}{36}$ c $\frac{3z^2}{2} - \frac{3}{2}$

Explanation	**Solution**
a Write the expression.	$p^2 - \frac{9}{49}$
Rewrite each term as something squared.	$= p^2 - \left(\frac{3}{7}\right)^2$
Factor using the DOTS formula.	$= \left(p - \frac{3}{7}\right)\left(p + \frac{3}{7}\right)$
b Write the expression.	$\frac{x^2}{25} - \frac{1}{36}$
Rewrite each term as something squared.	$= \left(\frac{x}{5}\right)^2 - \left(\frac{1}{6}\right)^2$
Factor using the DOTS formula.	$= \left(\frac{x}{5} - \frac{1}{6}\right)\left(\frac{x}{5} + \frac{1}{6}\right)$
c Write the expression.	$\frac{3z^2}{2} - \frac{3}{2}$
Factor out the GCF.	$= \frac{3}{2}(z^2 - 1)$
Factor using the DOTS formula.	$= \frac{3}{2}(z - 1)(z + 1)$

Worked Example 7

Factorise the following.

a $a^2 - 2$ b $7 - 4c^2$ c $3b^2 - 1$

Explanation		Solution
a	Write the expression.	$a^2 - 2$
	Rewrite each term as something squared.	$= a^2 - \left(\sqrt{2}\right)^2$
	Factor using the DOTS formula.	$= \left(a - \sqrt{2}\right)\left(a + \sqrt{2}\right)$
b	Write the expression.	$7 - 4c^2$
	Rewrite each term as something squared.	$= \left(\sqrt{7}\right)^2 - (2c)^2$
	Factor using the DOTS formula.	$= (\sqrt{7} - 2c)(\sqrt{7} + 2c)$
c	Write the expression.	$3b^2 - 1$
	Rewrite each term as something squared.	$= \left(\sqrt{3}b\right)^2 - 1^2$
	Factor using the DOTS formula.	$= \left(\sqrt{3}b - 1\right)\left(\sqrt{3}b + 1\right)$

If the number isn't a perfect square, you can still use DOTS.
Just write the number under a square root sign when factoring.

Your Turn To Try

1 What is the factored form of $x^2 - 16$?

A $(x+4)(x+4)$

B $(x-2)(x-8)$

C $(x-4)(x+4)$

D $(x+2)(x-8)$

2 What is the result of factoring $25 - y^2$?

A $(y-5)(y+5)$

B $(5-y)(5+y)$

C $(5-y)(5-y)$

D $(5+y)(5+y)$

3 If $36a^2 - 49$ is factored, what will it be?

A $(6a-7)(6a+7)$

B $(6a+7)(6a+7)$

C $(7-6a)(7+6a)$

D $(6a-7)(6a-7)$

4 What's the correct way to fully factor $4m^2 - 100$?

A $(2m-10)(2m+10)$

B $4(m-5)(m-5)$

C $2(m-5)(2m+10)$

D $4(m-5)(m+5)$

5 Which is the full factored form of $81 - 9x^2$.

A $-9(x-3)(x+3)$

B $-(9-3x)(9+3x)$

C $9(x-3)(x-3)$

D $9(x-3x)(x+3x)$

Question	1	2	3	4	5
Answer	C	B	A	D	A

Time To Practice

1 Factorise each of the following expressions.

a $x^2 - 1$ b $a^2 - 9$ c $b^2 - 121$

d $144 - x^2$ e $16 - a^2$ f $1 - 100b^2$

g $36x^2 - 1$ h $4a^2 - 25$ i $81b^2 - 16$

j $49 - 9x^2$ k $64 - 25a^2$ l $36b^2 - 1$

m $12x^2 - 48$ n $44 - 11a^2$ o $3b^2 - 27$

p $75 - 3x^2$ q $50a^2 - 8$ r $54 - 6b^2$

s $72x^2 - 98$ t $-27a^2 + 12$ u $125 - 20b^2$

2 Factor the following expressions.

a $(x + 2)^2 - 4$ b $(a + 5)^2 - 49$ c $(b - 2)^2 - 1$

d $(2x - 1)^2 - 1$ e $(4a - 5)^2 - 25$ f $(5b + 1)^2 - 64$

g $4 - (2x - 3)^2$ h $81 - (7a + 1)^2$ i $9 - (3b - 2)^2$

j $2(x + 5)^2 - 18$ k $-3(a - 2)^2 + 27$ l $2(1 - 5b)^2 - 98$

3 Fully factor each of the following expressions. Hint. You'll need to use surds.

a $x^2 - 7$ b $a^2 - 3$ c $5 - b^2$

d $4x^2 - 13$ e $9a^2 - 19$ f $3b^2 - 66$

g $15 - 5x^2$ h $4 - 2a^2$ i $36 - 48b^2$

j $(x + 3)^2 - 6$ k $15 - (a - 4)^2$ l $-2(b - 13)^2 + 14$

Challenge Yourself

1 Factorise the following expressions.

a $3ax^2 - 147a$

b $4m^2p - 36n^2p$

c $9x^2 - y^2z^2$

d $4x^2 - \frac{y^2}{121}$

e $\frac{x^2}{4} - \frac{9}{25}$

f $\frac{5x^2}{2} - \frac{5}{2}$

g $\frac{a^2b^2}{36} - \frac{c^2}{16}$

h $\frac{p^2q^2}{81} - \frac{m^2n^2}{100}$

The AC Method: Where A and C Get Together (B Joins In)

The AC Method.

It's for quadratics with three terms.

And it's the opposite of expanding double brackets.

A quadratic with three terms is in standard form.

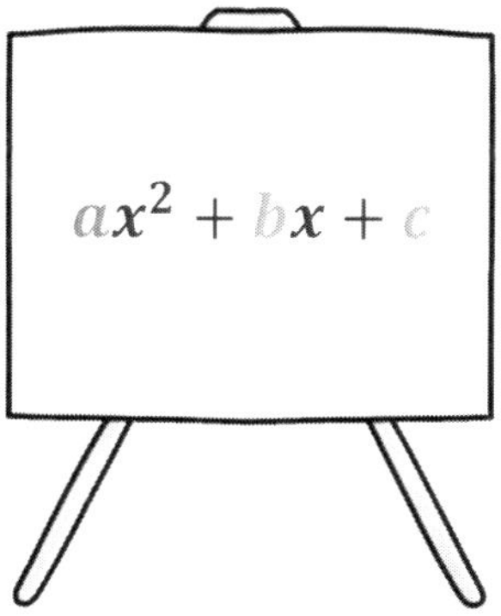

The number in front of the squared term is $\boldsymbol{a}$.

The number in front of the middle term is $\boldsymbol{b}$.

The number at the end is $\boldsymbol{c}$.

And $\boldsymbol{a}$, $\boldsymbol{b}$ and $\boldsymbol{c}$ include the sign in front of them.

AC, in the name means the a is multiplied by c.

Here's how it works.

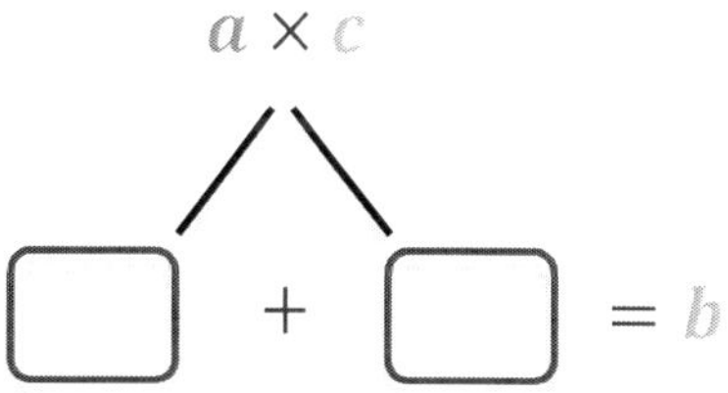

First, multiply whatever a and c are.

Then find two numbers that,

- Multiply to equal ac, **and**
- Add to equal b.

Once you find those two numbers.

Write two brackets.

In each bracket.

The first term is the variable.

The second term is one of those numbers with its sign.

That's pretty much it for the AC Method.

However, for this method to work, a can't be negative.

b and c can be positive or negative.

If a is negative.

Factor out -1 first.

Once you do that, a becomes positive.

Then factor what's left using this technique.

If there are a lot of pairs of numbers that multiply to ac.

You might need to write down some of them.

Then see which two add to b.

This is totally normal.

Everyone does it.

It's part of the process.

What matters is finding the right pair and getting the right signs.

Here are the steps, broken down.

1) Make sure the quadratic is in standard form
2) Take out the GCF (if there's one)
3) Multiply a and c
4) Find two numbers that multiply to equal ac and add to equal b
5) In each bracket, write the variable and one of these numbers with its sign

Worked Example 1

Factorise the following.

a $x^2 + 5x + 6$ b $x^2 + 8x + 12$ c $x^2 - 2x - 15$

Explanation		Solution
a	Write the expression.	$x^2 + 5x + 6$
	Multiply a and c and write the value of b.	$ac = 6 \quad b = 5$
	Find two numbers that multiply to ac and add to b.	$2 \times 3 = 6 \rightarrow ac$ $2 + 3 = 5 \rightarrow b$
	Write the factors using these numbers.	$= (x + 2)(x + 3)$
b	Write the expression.	$x^2 + 8x + 12$
	Multiply a and c and write the value of b.	$ac = 12, \quad b = 8$
	Find two numbers that multiply to ac and add to b.	$2 \times 6 = 12 \rightarrow ac$ $2 + 6 = 8 \rightarrow b$
	Write the factors using these numbers.	$= (x + 2)(x + 6)$
c	Write the expression.	$x^2 - 2x - 15$
	Multiply a and c and write the value of b.	$ac = -15 \quad b = -2$
	Find two numbers that multiply to ac and add to b.	$-5 \times 3 = -15 \rightarrow ac$ $-5 + 3 = -2 \rightarrow b$
	Write the factors using these numbers.	$= (x - 5)(x + 3)$

Worked Example 2

Factorise the following.

a $\mathbf{p^2 - 17p + 30}$ b $\mathbf{m^2 - 6m - 27}$

Explanation	**Solution**
a Write the expression.	$p^2 - 17p + 30$
Multiply a and c and write the value of b.	$ac = 30 \quad b = -17$
Find two numbers that multiply to ac and add to b.	$\mathbf{-15 \times -2 = 30 \rightarrow ac}$ $\mathbf{-15 + (-2) = -17 \rightarrow b}$
Write the factors using these numbers.	$= (p - 15)(p - 2)$
b Write the expression.	$m^2 - 6m - 27$
Multiply a and c and write the value of b.	$ac = -27 \quad b = -6$
Find two numbers that multiply to ac and add to b.	$\mathbf{-9 \times 3 = -27 \rightarrow ac}$ $\mathbf{-9 + 3 = -6 \rightarrow b}$
Write the factors using these numbers.	$= (x - 9)(x + 3)$

If a is 1 (nothing in front of the squared term), find two numbers that multiply to equal c.

Worked Example 3

Factorise the following.

a $-x^2 + 9x - 14$

b $2x^2 + 10x - 12$

Explanation	Solution
a Write the expression.	$-x^2 + 9x - 14$
Factor out the negative.	$= -(x^2 - 9x + 14)$
Multiply a and c and write the value of b.	$ac = 14 \quad b = -9$
Find two numbers that multiply to ac and add to b.	$-7 \times -2 = 14 \rightarrow ac$ $-7 + (-2) = -9 \rightarrow b$
Write the factors using these numbers.	$= -(x - 7)(x - 2)$
b Write the expression.	$2x^2 + 10x - 12$
Factor out the GCF.	$= 2(x^2 + 5x - 6)$
Multiply a and c and write the value of b.	$ac = -6 \quad b = 5$
Find two numbers that multiply to ac and add to b.	$-1 \times 6 = -6 \rightarrow ac$ $-1 + 6 = 5 \rightarrow b$
Write the factors using these numbers.	$= 2(x - 1)(x + 6)$

If the quadratic isn't written in standard form, rewrite it so it is.

Worked Example 4

Factorise the following.

a $\mathbf{72 - 17x + x^2}$

b $\mathbf{51x - 3x^2 - 198}$

Explanation		**Solution**
a	Write the expression in standard form.	$x^2 - 17x + 72$
	Multiply a and c and write the value of b.	$ac = 72 \quad b = -17$
	Find two numbers that multiply to ac and add to b.	$\mathbf{-9 \times -8 = 72} \rightarrow \boldsymbol{ac}$ $\mathbf{-9 + (-8) = -17} \rightarrow \boldsymbol{b}$
	Write the factors using these numbers.	$= (x - 9)(x - 8)$
b	Write the expression in standard form.	$-3x^2 + 51x - 198$
	Factor out the GCF.	$= -3(x^2 - 17x + 66)$
	Multiply a and c and write the value of b.	$ac = 66 \quad b = -17$
	Find two numbers that multiply to ac and add to b.	$\mathbf{-11 \times -6 = 66} \rightarrow \boldsymbol{ac}$ $\mathbf{-11 + (-6) = -17} \rightarrow \boldsymbol{b}$
	Write the factors using these numbers.	$= -3(x - 11)(x - 6)$

Always check to see if a can be factored out first.
Whether its positive or negative.

Worked Example 5

Factorise $\frac{1}{4}x^2 - \frac{1}{2}x - 6$

Explanation	Solution
Write the expression in standard form.	$\frac{1}{4}x^2 - \frac{1}{2}x - 6$
Factor out the GCF.	$= \frac{1}{4}(x^2 - 2x - 24)$
Multiply a and c and write the value of b.	$ac = -24 \quad b = -2$
Find two numbers that multiply to ac and add to b.	$-6 \times 4 = -24 \rightarrow ac$ $-6 + 4 = -2 \rightarrow b$
Write the factors using these numbers.	$= \frac{1}{4}(x - 6)(x + 4)$

When factoring out a fraction, multiply each term by the its reciprocal.
Put the new expression inside brackets.

Worked Example 6

Factorise $(x+8)^2 - (x+8) - 2$

Explanation	**Solution**
Write the expression.	$(x+8)^2 - (x+8) - 2$
Use the substitution method.	Let $u = x + 8$
Rewrite quadratic in terms of u.	$= u^2 - u - 2$
Multiply a and c and write the value of b.	$ac = -2 \quad b = -1$
Find two numbers that multiply to ac and add to b.	$\mathbf{-2 \times 1 = -2 \rightarrow ac}$ $\mathbf{-2 + 1 = -1 \rightarrow b}$
Write the factors using these numbers.	$= (u-2)(u+1)$
Replace u with original expression.	$= (x+8-2)(x+8+1)$
Simplify.	$= (x+6)(x+9)$

The AC Method 2: When 'a' is Not One (This'll Be Fun)

When the value of a is some other number.

It's not 1 or -1.

Still use the AC Method to factor.

Except for one thing.

There's a little trick you need to do.

Once you find the two numbers that multiply to ac and add to b.

Divide each of those numbers by whatever a is.

And write them as a fraction.

If a fraction can be simplified.

Simplify it.

But keep it as a fraction.

If it doesn't simplify.

Leave it as a fraction.

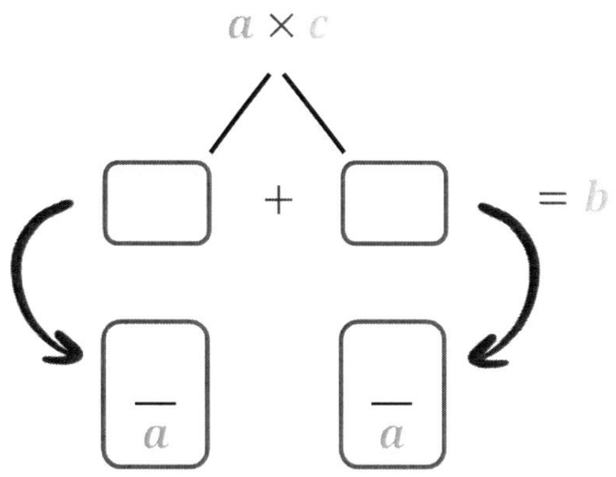

Then write two brackets.

The first term in each bracket will be the variable.

Now, here's the trick.

Using the fractions you end up with.

The denominator is the number that goes in front of the variable.

The numerator is the number added or subtracted to it.

Here are the steps to follow.

1) Write the quadratic in standard form
2) Take out the GCF (if there is one)
3) Multiply a and c and write down the value of b.
4) Find two numbers that multiply to equal ac and add to equal b
    i. Divide each of these numbers by a and write them as a fraction
    ii. Simplify the fractions, if needed, but keep them as fractions
5) Write two brackets with a variable term and a number term in each
    i. The denominator goes in front of the variable
    ii. The numerator is the number being added or subtracted

Worked Example 7

Factorise the following.

a $2x^2 + 13x + 6$

b $2x^2 + 3x - 9$

Explanation	Solution
a Write the expression in standard form.	$2x^2 + 13x + 6$
Multiply a and c and write the value of b.	$ac = 12 \quad b = 13$
Find two numbers that multiply to ac and add to b.	$1 \times 12 = 12 \rightarrow ac$ $1 + 12 = 13 \rightarrow b$
Divide numbers found by a and simplify.	$\frac{1}{2} = \frac{1}{2}$ *and* $\frac{12}{2} = \frac{6}{1}$
Write the factors using the numerator and denominator.	$= (2x + 1)(x + 6)$
b Write the expression in standard form.	$2x^2 + 3x - 9$
Multiply a and c and write the value of b.	$ac = -18 \quad b = 3$
Find two numbers that multiply to ac and add to b.	$-3 \times 6 = -18 \rightarrow ac$ $-3 + 6 = 3 \rightarrow b$
Divide numbers found by a and simplify.	$\frac{-3}{2} = \frac{-3}{2}$ *and* $\frac{6}{2} = \frac{3}{1}$
Write the factors using the numerator and denominator.	$= (2x - 3)(x + 3)$

Worked Example 8

Factorise the following.

a $\mathbf{3x^2 - 8x + 4}$

b $\mathbf{3x^2 - 2x - 5}$

Explanation		Solution
a	Write the expression in standard form.	$3x^2 - 8x + 4$
	Multiply a and c and write the value of b.	$ac = 12 \quad b = -8$
	Find two numbers that multiply to ac and add to b.	$-6 \times -2 = 12 \rightarrow ac$ $-6 + (-2) = -8 \rightarrow b$
	Divide numbers found by a and simplify.	$\frac{-6}{3} = \frac{-2}{1}$ *and* $\frac{-2}{3} = \frac{-2}{3}$
	Write the factors using the numerator and denominator.	$= (x - 2)(3x - 2)$
b	Write the expression in standard form.	$3x^2 - 2x - 5$
	Multiply a and c and write the value of b.	$ac = -15 \quad b = -2$
	Find two numbers that multiply to ac and add to b.	$-5 \times 3 = -15 \rightarrow ac$ $-5 + 3 = -2 \rightarrow b$
	Divide numbers found by a and simplify.	$\frac{-5}{3} = \frac{-5}{3}$ *and* $\frac{3}{3} = \frac{1}{1}$
	Write the factors using the numerator and denominator.	$= (3x - 5)(x + 1)$

Worked Example 9

Factorise the following.

a $\mathbf{4x^2 + 3x - 10}$

b $\mathbf{12x^2 + 16x - 3}$

Explanation	Solution
a Write the expression.	$4x^2 + 3x - 10$
Multiply a and c and write the value of b.	$ac = -40 \quad b = 3$
Find two numbers that multiply to ac and add to b.	$-5 \times 8 = -40 \rightarrow ac$ $-5 + 8 = 3 \rightarrow b$
Divide numbers found by a and simplify.	$\frac{-5}{4} = \frac{-5}{4}$ *and* $\frac{8}{4} = \frac{2}{1}$
Write the factors using the numerator and denominator.	$= (4x - 5)(x + 2)$
b Write the expression in standard form.	$12x^2 + 16x - 3$
Multiply a and c and write the value of b.	$ac = -36 \quad b = 16$
Find two numbers that multiply to ac and add to b.	$18 \times -2 = 36 \rightarrow ac$ $18 + (-2) = 16 \rightarrow b$
Divide numbers found by a and simplify.	$\frac{18}{12} = \frac{3}{2}$ *and* $\frac{-2}{12} = \frac{-1}{6}$
Write the factors using the numerator and denominator.	$= (2x - 3)(6x - 1)$

Worked Example 10

Factorise $\frac{5}{2}x^2 - \frac{11}{2}x + 1$

Explanation	Solution
Write the expression.	$\frac{5}{2}x^2 - \frac{11}{2}x + 1$
Factor out the GCF.	$= \frac{1}{2}(5x^2 - 11x + 2)$
Multiply a and c and write the value of b.	$ac = 10 \quad b = -11$
Find two numbers that multiply to ac and add to b.	$-10 \times -1 = 10 \rightarrow ac$ $-10 + (-1) = -11 \rightarrow b$
Divide numbers found by a and simplify.	$\frac{-10}{5} = \frac{-2}{1}$ and $\frac{-1}{5} = \frac{-1}{5}$
Write the factors using the numerator and denominator.	$= \frac{1}{2}(x - 2)(5x - 1)$

Worked Example 11

Factorise $4 - \frac{7}{3}x^2 - \frac{40}{3}x$

Explanation	**Solution**
Write the expression in standard form.	$-\frac{7}{3}x^2 - \frac{40}{3}x + 4$
Factor out the GCF.	$= -\frac{1}{3}(7x^2 + 40x - 12)$
Multiply a and c and write the value of b.	$ac = -84 \quad b = 40$
Find two numbers that multiply to ac and add to b.	$-2 \times 42 = -84 \rightarrow ac$ $-2 + 42 = 40 \rightarrow b$
Divide numbers found by a and simplify.	$\frac{-2}{7} = \frac{-2}{7}$ *and* $\frac{42}{7} = \frac{6}{1}$
Write the factors using the numerator and denominator.	$= -\frac{1}{3}(7x - 2)(x + 6)$

When you have two different fractions, the smaller one (i.e. the one with the bigger denominator) can be factored out as the GCF.

Worked Example 12

$4x^2 + px + 6$ is factorised as $(2x + 3)(2x + q)$. Find the value of p.

Explanation	Solution
Write the expression.	$4x^2 + px + 6$
Write in factored form.	$= (2x + 3)(2x + q)$
Solve for q.	$3q = 6$ $q = 2$
Rewrite in factored form.	$= (2x + 3)(2x + 2)$
Expand brackets.	$= 4x^2 + 4x + 6x + 6$
	$= 4x^2 + 10x + 6$
Compare expressions.	$4x^2 + 10x + 6 = 4x^2 + px + 6$
	$\therefore p = 10$

Expanding brackets in the form $(x + p)(x + q)$. The number $p \times q$ is the c value in the quadratic.

Worked Example 13

Given $6x^2 + kx + 5$ can be factorised, and one factor is $(2x + 1)$. Find the value of k, where k is a constant.

Explanation	Solution
Write the expression.	$6x^2 + kx + 5$
Write in factored form with given factor.	$= (2x + 1)(px + q)$
Solve for p and q.	$1 \times q = 5$ $2x \times px = 6x^2$
	$q = 5$ $2px^2 = 6x^2$
	$2p = 6$
	$p = 3$
Rewrite in factored form.	$= (2x + 1)(3x + 5)$
Expand brackets.	$= 6x^2 + 10x + 3x + 5$
	$= 6x^2 + 13x + 5$
Compare expressions.	$6x^2 + 13x + 5 = 6x^2 + kx + 5$
	$\therefore k = 13$

Your Turn To Try

1 **What is the first step to factor $x^2 + 5x - 24$ using the AC Method?**

A Factor out the GCF

B Find the factors of the middle term

C Rewrite in standard form

D Multiply the a and c values

2 **For the quadratic $3x^2 + 4x - 4$ what is the value of AC?**

A -12

B -4

C 12

D 4

3 **If $a^2 - 7a + 10$ is factored, what will it be?**

A $(a-5)(a+2)$

B $(a+5)(a+2)$

C $(a-5)(a-2)$

D $(a+5)(a-2)$

4 **Given $3p^2 - 8p - 3$, which pair of numbers multiplies to the AC value and adds to the b value?**

A 3 and -3

B -9 and 1

C -1 and 9

D 1 and 9

5 **What's the fully factored form of $2m^2 + 5m + 3$?**

A $(m+2)(2m+1)$

B $(m+2)(2m+3)$

C $(m+1)(2m+3)$

D $(m+3)(2m+1)$

Question	1	2	3	4	5
Answer	D	A	C	B	C

Time To Practice

1 Factorise each of the following quadratics.

a $x^2 + 4x + 3$ b $a^2 - 7a + 10$ c $b^2 - 4b - 12$

d $x^2 - 8x + 15$ e $a^2 - 9a + 20$ f $b^2 + b - 110$

g $-x^2 + 6x - 8$ h $-a^2 - 3a + 10$ i $-b^2 - 5b + 24$

j $4x^2 + 20x + 16$ k $6a^2 - 12a - 18$ l $2b^2 - 26b + 84$

2 Factor the following quadratics.

a $12 + 18x + 6x^2$ b $18 - 3a^2 - 15a$ c $14b - 24 - 2b^2$

d $9x^2 - 27x + 18$ e $-5a^2 - 25a - 30$ f $28b - 48 - 4b^2$

g $15x^2 + 45x + 30$ h $27a - 18 - 9a^2$ i $48b - 6b^2 - 42$

3 Factor each of the following quadratics.

a $2x^2 + 5x + 3$ b $3a^2 - 8a + 4$ c $4b^2 + 4b - 3$

d $6x^2 + 17x + 5$ e $6a^2 + 7a - 5$ f $8b^2 + 14b + 3$

g $12x^2 + 5x - 3$ h $-9a^2 + 3a + 2$ i $-6b^2 + 13b + 5$

j $-3x^2 + 11x - 10$ k $-2a^2 - 15a - 18$ l $-4b^2 + 7b + 2$

4 Factor each of the following quadratics.

a $6x^2 - 10x - 4$ b $-6a^2 - 9a + 6$ c $8b^2 - 4b - 12$

d $-15x^2 - 25x + 10$ e $14a^2 + 49a + 21$ f $-12b^2 + 54b - 60$

g $44x^2 - 132x - 77$ h $30a^2 + 25a - 30$ i $-16b^2 + 20b - 6$

Time To Practice

5 Factorise each of the following quadratics.

a $12x + 18x^2 - 30$

b $20 - 6a^2 + 2a$

c $48 - 20b - 8b^2$

d $21x^2 - 21 - 56x$

e $30 - 54a - 12a^2$

f $32 + 16b - 6b^2$

g $63x + 18x^2 - 36$

h $18 - 27b^2 - 45b$

i $8 - 8a - 96a^2$

Challenge Yourself

1 Given the quadratic $2x^2 - 11x + 5$ is factored as $(2x - 1)(x + k)$. What is the value of k?

2 Given the quadratic $x^2 + (k - 5)x - k$ is factored as $(x - 2)(x + 3)$. What is the value of k?

3 The quadratic $x^2 + kx + 3$ can be factored as $(x + 3)(x + m)$. What is the value of m and k?

4 If $3x^2 - kx + 12 = (3x + 4)(x + p)$, what is the value of p and k?

5 For what values of k, where k is a positive whole number, can $3x^2 + kx + 5$ be factored?

Completing the Square: A Bit of Perfect Squares, A Bit of DOTS

The AC Method doesn't work all the time.

It only works when two whole numbers multiply to ac and add to b.

Sometimes there aren't any two numbers that work.

To factor those, use Completing the Square.

For this technique you need to know how to factor perfect squares.

And the difference of two squares (DOTS).

Because it uses both together.

In the one question.

Here's how it works.

The quadratic has to be in standard form.

It's then rewritten in a specific way.

In a way where it can be split into two parts.

One part will always be a perfect square.

Which gets factored using the perfect square formula.

After it's factored.

The quadratic will then look like a difference of two squares.

From there.

It's then factored again using the DOTS formula.

That's the gist of it.

Now, here's how you do it.

Look at the number b.

Half that number, then square it.

Add this squared number after the middle term, and subtract it straight after.

If the original quadratic had a number at the end.

Write that number too.

The first three terms now become a perfect square.

You can put brackets around them to make it easier to see.

Factor this part separately, and combine the leftover numbers.

The quadratic will now look like a DOTS.

Factor it using the DOTS formula.

So why bother with this technique?

If quadratics don't factor nicely using whole numbers.

This is the way to go.

It always works.

Later in math.

Completing the square becomes really handy.

Like when sketching quadratics.

Having it in DOTS form tells you heaps about the graph.

Just by looking at it.

The DOTS form also makes quadratics easier to solve.

Here are the steps.

1) Write the quadratic in standard form
2) Factor out the GCF (if needed)
3) Half the b number $\left(\frac{b}{2}\right)$, then square it $\left(\frac{b}{2}\right)^2$
4) Add and subtract this squared number $\left(\frac{b}{2}\right)^2$ right after the middle term
5) Group the first three terms inside brackets
6) Combine the leftover numbers
7) Factor the perfect square part
8) Factor using the difference of two squares
9) Simplify if needed

This technique also works for quadratics that factor with whole numbers.
Even some of the two-term ones.
But it's mainly for three-term quadratics that don't factor nicely.

The issue with it is that it has a lot of steps.
If a quadratic can be factored using the other methods.
It's quicker to use them.

We'll start with ones that factor nicely with whole numbers.
Just to get the process.
Then we'll work on the ones that don't.

Worked Example 1

Factorise by completing the square: $x^2 + 10x$.

Explanation	Solution
Write the expression.	$x^2 + 10x$
Half the b number and square it.	$\rightarrow \left(\frac{10}{2}\right)^2 = (5)^2$
Add and then subtract this squared number after the middle term.	$= x^2 + 10x + (5)^2 - (5)^2$
Group the first three terms inside brackets.	$= [x^2 + 10x + (5)^2] - 25$
Factor the perfect square part.	$= (x+5)^2 - 25$
Factor using the difference of two squares method.	$= (x+5-5)(x+5+5)$
Simplify.	$= x(x+10)$

Use a different color for $\left(\frac{b}{2}\right)^2$ when adding and subtracting.
It makes the perfect squares stand out.

Worked Example 2

Factorise by completing the square: $x^2 - 4x - 12$.

Explanation	Solution
Write the expression.	$x^2 - 4x - 12$
Half the b number and square it.	$\rightarrow \left(\frac{4}{2}\right)^2 = (2)^2$
Add and then subtract this squared number after the middle term.	$= x^2 - 4x + (2)^2 - (2)^2 - 12$
Group the first three terms inside brackets.	$= [x^2 - 4x + (2)^2] - 4 - 12$
Factor the perfect square part and combine the leftover numbers.	$= (x - 2)^2 - 16$
Factor using the difference of two squares method.	$= (x - 2 - 4)(x - 2 + 4)$
Simplify.	$= (x - 6)(x + 2)$

Always add the $\left(\frac{b}{2}\right)^2$ first.
This makes the first three terms turn into a perfect square.

Worked Example 3

Fully factorise by completing the square: $x^2 + 6x - 1$.

Explanation	Solution
Write the expression.	$x^2 + 6x - 1$
Half the b number and square it.	$\rightarrow \left(\frac{6}{2}\right)^2 = (3)^2$
Add and then subtract this squared number after the middle term.	$= x^2 + 6x + (3)^2 - (3)^2 - 1$
Group the first three terms inside brackets.	$= [x^2 + 6x + (3)^2] - 9 - 1$
Factor the perfect square part and combine the leftover numbers.	$= (x + 3)^2 - 10$
Factor using the difference of two squares method.	$= \left(x + 3 - \sqrt{10}\right)\left(x + 3 + \sqrt{10}\right)$

Squaring the b value instead of halving it first and then squaring it. ⚠

Worked Example 4

Fully factorise by completing the square: $x^2 - 8x + 10$.

Explanation	Solution
Write the expression.	$x^2 - 8x + 10$
Half the b number and square it.	$\rightarrow \left(\frac{8}{2}\right)^2 = (4)^2$
Add and then subtract this squared number after the middle term.	$= x^2 - 8x + (4)^2 - (4)^2 + 10$
Group the first three terms inside brackets.	$= [x^2 - 8x + (4)^2] - 16 + 10$
Factor the perfect square part and combine the leftover numbers.	$= (x - 4)^2 - 6$
Factor using the difference of two squares method.	$= \left(x - 4 - \sqrt{6}\right)\left(x - 4 + \sqrt{6}\right)$

You can ignore the negative sign in front of b. Squaring cancels it out anyway.

Worked Example 5

Fully factorise by completing the square: $-x^2 - 2x + 7$

Explanation	**Solution**
Write the expression.	$-x^2 - 2x + 7$
Take out a negative as a factor.	$= -(x^2 + 2x - 7)$
Half the b number and square it.	$\rightarrow \left(\frac{2}{2}\right)^2 \rightarrow (1)^2 \rightarrow 1$
Add and then subtract this squared number after the middle term.	$= -(x^2 + 2x + (1)^2 - (1)^2 - 7)$
Group the first three terms inside brackets.	$= -([x^2 + 2x + (1)^2] - 1 - 7)$
Factor the perfect square part and combine the leftover numbers.	$= -((x + 1)^2 - 8)$
Factor using the difference of two squares method.	$= -\left(x + 1 - \sqrt{8}\right)\left(x + 1 + \sqrt{8}\right)$
Simply the surds.	$= -\left(x + 1 - 2\sqrt{2}\right)\left(x + 1 + 2\sqrt{2}\right)$

If the coefficient of x^2 is negative, factor out the negative.
You always want that coefficient to be positive 1.

Worked Example 6

Fully factorise by completing the square: $x^2 - 7x + 8$.

Explanation	**Solution**
Write the expression.	$x^2 - 7x + 8$
Half the b number and square it.	$\rightarrow \left(\frac{7}{2}\right)^2$
Add and then subtract this squared number after the middle term.	$= x^2 - 7x + \left(\frac{7}{2}\right)^2 - \left(\frac{7}{2}\right)^2 + 8$
Group the first three terms inside brackets.	$= \left[x^2 - 7x + \left(\frac{7}{2}\right)^2\right] - \frac{49}{4} + 8$
Factor the perfect square part and combine the leftover numbers.	$= \left(x - \frac{7}{2}\right)^2 - \frac{49}{4} + \frac{32}{4}$ $= \left(x - \frac{7}{2}\right)^2 - \frac{17}{4}$
Factor using the difference of two squares method.	$= \left(x - \frac{7}{2} - \frac{\sqrt{17}}{2}\right)\left(x - \frac{7}{2} + \frac{\sqrt{17}}{2}\right)$
Simplify by combining fractions.	$= \left(x - \frac{7 - \sqrt{17}}{2}\right)\left(x - \frac{7 + \sqrt{17}}{2}\right)$

Worked Example 7

Factorise by completing the square: $x^2 - 2x + 8$

Explanation	Solution
Write the expression.	$x^2 - 2x + 8$
Half the b number and square it.	$\rightarrow \left(\frac{2}{2}\right)^2 = (1)^2$
Add and then subtract this squared number after the middle term.	$= x^2 - 2x + (1)^2 - (1)^2 + 8$
Group the first three terms inside brackets.	$= [x^2 - 2x + (1)^2] - 1^2 + 8$
Factor the perfect square part and combine the leftover numbers.	$= (x - 1)^2 + 7$
(Can't factor any further as we don't have difference of two squares).	

Sometimes, it won't turn into a difference of two squares, and that's fine. Just leave it as a squared bracket form plus the number.

Completing the Square 2: When 'a' is Not One (This Won't Be Fun)

Completing the square only works when a is 1.

Anything else?

This technique won't work.

If a is not 1.

First thing.

Factor it out.

Then follow the same steps to factor what's left inside the brackets.

It does get a bit messy.

There'll be brackets within brackets.

So be careful with what goes where.

When you factor out a first.

This usually turns the middle and third terms into fractions.

You'll be halving them.

Squaring them.

Adding and subtracting them.

Even taking their square roots.

Which means you need to be good with fractions.

And surds too.

One small mistake can mess everything up.

One more thing.

At the end.

See if you can simplify each factor by combining the number parts.

Write them as one number term.

Keeps things clean.

Worked Example 8

Fully factorise by completing the square: $2x^2 - 8x - 6$

Explanation	Solution
Write the expression.	$2x^2 - 8x - 6$
Factor out a.	$= 2(x^2 - 4x - 3)$
Half the b number and square it.	$\rightarrow \left(\frac{4}{2}\right)^2 = (2)^2$
Add and then subtract this squared number after the middle term.	$= 2(x^2 - 4x + (2)^2 - (2)^2 - 3)$
Group the first three terms inside brackets.	$= 2([x^2 - 4x + (2)^2] - 4 - 3)$
Factor the perfect square part and combine the leftover numbers.	$= 2((x-2)^2 - 7)$
Factor using the difference of two squares method.	$= 2(x - 2 - \sqrt{7})(x - 2 + \sqrt{7})$

If asked to 'factorise,' stop once you have the squared bracket form.
If asked to 'fully factorise,' go one step further and factor it into two brackets.

Worked Example 9

Fully factorise by completing the square: $24x - 3x^2 - 15$

Explanation	Solution
Write the expression.	$-3x^2 + 24x - 15$
Factor out a.	$= -3(x^2 - 8x + 5)$
Half the b number and square it.	$\rightarrow \left(\frac{8}{2}\right)^2 = (4)^2$
Add and then subtract this squared number after the middle term.	$= -3(x^2 - 8x + (4)^2 - (4)^2 + 5)$
Group the first three terms inside brackets.	$= -3([x^2 - 8x + (4)^2] - 16 + 5)$
Factor the perfect square part and combine the leftover numbers.	$= -3((x-4)^2 - 11)$
Factor using the difference of two squares method.	$= -3(x - 4 - \sqrt{11})(x - 4 + \sqrt{11})$

Worked Example 10

Fully factorise by completing the square: $2x^2 - 3x - 7$

Explanation	Solution
Write the expression.	$2x^2 - 3x - 7$
Factor out a.	$= 2\left(x^2 - \frac{3}{2}x - \frac{7}{2}\right)$
Half the b number and square it.	$\rightarrow \left(\frac{3}{4}\right)^2$
Add and then subtract this squared number after the middle term.	$= 2\left(x^2 - \frac{3}{2}x + \left(\frac{3}{4}\right)^2 - \left(\frac{3}{4}\right)^2 - \frac{7}{2}\right)$
Group the first three terms inside brackets.	$= 2\left(\left[x^2 - \frac{3}{2}x + \left(\frac{3}{4}\right)^2\right] - \frac{9}{16} - \frac{56}{16}\right)$
Factor the perfect square part and combine the leftover numbers.	$= 2\left(\left(x - \frac{3}{4}\right)^2 - \frac{65}{16}\right)$
Factor using the difference of two squares method.	$= 2\left(x - \frac{3}{4} - \frac{\sqrt{65}}{4}\right)\left(x + \frac{3}{4} + \frac{\sqrt{65}}{4}\right)$
Simply by combining fractions.	$= 2\left(x - \frac{3 - \sqrt{65}}{4}\right)\left(x - \frac{3 + \sqrt{65}}{4}\right)$

To half a fraction, multiply its denominator by 2 and leave the numerator as it is.

Worked Example 11

Fully factorise by completing the square: $\frac{5}{2} - 4x^2 - 2x$

Explanation	Solution
Write the expression.	$-4x^2 - 2x + \frac{5}{2}$
Factor out a.	$= -4\left(x^2 + \frac{1}{2}x - \frac{5}{8}\right)$
Half the b number and square it.	$\rightarrow \left(\frac{1}{4}\right)^2$
Add and then subtract this squared number after the middle term.	$= -4\left(x^2 + \frac{1}{2}x + \left(\frac{1}{4}\right)^2 - \left(\frac{1}{4}\right)^2 - \frac{5}{8}\right)$
Group the first three terms inside brackets.	$= -4\left(\left[x^2 + \frac{1}{2}x + \left(\frac{1}{4}\right)^2\right] - \frac{1}{16} - \frac{10}{16}\right)$
Factor the perfect square part and combine the leftover numbers.	$= -4\left(\left(x + \frac{1}{4}\right)^2 - \frac{11}{16}\right)$
Factor using the difference of two squares method.	$= -4\left(x + \frac{1}{4} - \frac{\sqrt{11}}{4}\right)\left(x + \frac{1}{4} + \frac{\sqrt{11}}{4}\right)$
Simply by combining fractions.	$= -4\left(x + \frac{1 - \sqrt{11}}{4}\right)\left(x + \frac{1 + \sqrt{11}}{4}\right)$

Worked Example 12

Fully factorise by completing the square: $3x^2 + 12x - 2$

Explanation	Solution
Write the expression.	$3x^2 + 12x - 2$
Factor out a.	$= 3\left(x^2 + 4x - \frac{2}{3}\right)$
Half the b number and square it.	$\left(\frac{4}{2}\right)^2 \rightarrow (2)^2$
Add and then subtract this squared number after the middle term.	$= 3\left(x^2 + 4x + (2)^2 - (2)^2 - \frac{2}{3}\right)$
Group the first three terms inside brackets.	$= 3\left([x^2 + 4x + (2)^2] - 4 - \frac{2}{3}\right)$
Factor the perfect square part and combine the leftover numbers.	$= 3\left((x+2)^2 - \frac{12}{3} - \frac{2}{3}\right)$
	$= 3\left((x+2)^2 - \frac{14}{3}\right)$
Factor using the difference of two squares method.	$= 3\left(x + 2 - \sqrt{\frac{14}{3}}\right)\left(x + 2 + \sqrt{\frac{14}{3}}\right)$
Rationalize the surd.	$\sqrt{\frac{14}{3}} = \frac{\sqrt{14}}{\sqrt{3}} \rightarrow \frac{\sqrt{14}}{\sqrt{3}} \times \frac{\sqrt{3}}{\sqrt{3}} = \frac{\sqrt{42}}{3}$
Rewrite factors in rationalized form.	$= 3\left(x + 2 - \frac{\sqrt{42}}{3}\right)\left(x + 2 + \frac{\sqrt{42}}{3}\right)$
Simply by combining fractions.	$= 3\left(x + \frac{6 - \sqrt{42}}{3}\right)\left(x + \frac{6 + \sqrt{42}}{3}\right)$

Worked Example 13

Find the value of k for the following.

a **The quadratic $x^2 + kx + 9$ can be written in the form $(x + 4)^2 - 7$**

b **The quadratic $3x^2 - kx + 17$ can be written in the form $3(x - 2)^2 + 5$**

Explanation		Solution
a	Write the expressions.	$(x + 4)^2 - 7$
	Expand brackets.	$= x^2 + 8x + 16 - 7$
	Simplify.	$= x^2 + 8x + 9$
	Let the two expressions equal.	$x^2 + 8x + 9 = x^2 + kx + 9$
	Match the coefficients.	$\therefore\ k = 8$
b	Write the expressions.	$3(x - 2)^2 + 5$
	Expand brackets.	$= 3(x^2 - 4x + 4) + 5$ $= 3x^2 - 12x + 12 + 5$
	Simplify.	$= 3x^2 - 12x + 17$
	Let the two expressions equal.	$3x^2 - 12x + 17 = 3x^2 - kx + 17$
	Match the coefficients.	$\therefore\ k = 12$

Your Turn To Try

1 **Which step is NOT part of the complete the square method?**

A Write in standard form
B Multiply by coefficient of x^2
C Halve & square b value
D Factor as perfect squares

2 **In the expression $x^2 + 6x - 5$ what number would you add and subtract to complete the square?**

A 3
B 12
C 6
D 9

3 **After completing the square $a^2 + 8a - 5$ becomes?**

A $(a+4)^2 - 21$
B $(a+4)^2 - 9$
C $(a+2)^2 - 21$
D $(a+2)^2 - 9$

4 **If you're completing the square for $2p^2 + 12p + 7$, what's the first step?**

A Halve 12 and square it
B Add and subtract 36
C Factor out 2
D Factor out 12

5 **Which expression is equivalent to $2x^2 + 4x$ after completing the square?**

A $2(x^2 + 2x)$
B $2(x+1)^2$
C $2(x+1)^2 - 1$
D $2(x+1)^2 - 2$

Question	1	2	3	4	5
Answer	B	D	A	C	D

Time To Practice

1 Use complete the square method to factorise the following quadratics.

a $x^2 + 12x$ b $a^2 - 10a$ c $b^2 - 2b$

d $x^2 + 8x$ e $a^2 - 4a$ f $b^2 - 16b$

g $x^2 + 9x$ h $a^2 - 7a$ i $b^2 + 11b$

j $x^2 - x$ k $a^2 - 3a$ l $b^2 - 5b$

2 Fully factorise the following quadratics by completing the square.

a $x^2 + 4x + 1$ b $a^2 + 6a + 2$ c $b^2 + 2b - 4$

d $x^2 + 10x - 4$ e $a^2 - 8a + 13$ f $b^2 - 12b + 10$

g $x^2 - 4x - 3$ h $a^2 - 8a - 5$ i $b^2 + 14b + 6$

j $7 + 4x - x^2$ k $2 - 2a - a^2$ l $10b - 12 - b^2$

3 Fully factorise the following quadratics by completing the square.

a $2x^2 + 4x - 4$ b $4a^2 - 8a - 20$ c $5b^2 + 20b - 5$

d $3x^2 - 12x - 39$ e $5a^2 - 30a + 10$ f $2b^2 - 28b + 94$

g $-3x^2 - 30x - 18$ h $14 + 8a - 2a^2$ i $8b + 16 - 4b^2$

4 Fully factorise the following quadratics by completing the square.

a $3x^2 - 4x - 1$ b $5a^2 - 2a - 1$ c $8b^2 - 12b - 1$

d $-2x^2 - 10x - 11$ e $3a^2 - 2a - 2$ f $2b + 2 - 3b^2$

g $3 - 2x - 3x^2$ h $18a - 9a^2 - 4$ i $4b^2 + 16b + 3$

Challenge Yourself

1 Fully factorise the following quadratics using the complete the square method.

a $x^2 - 4x + \frac{3}{2}$

b $-x^2 + 6x - \frac{16}{3}$

c $-3x^2 - 4x + \frac{13}{6}$

d $2a^2 + 2a - \frac{25}{6}$

e $3a^2 - 2a - \frac{4}{15}$

f $-6x^2 + 2x + \frac{1}{3}$

2 For what value of c will this quadratic $x^2 + 8x + c$ become a perfect square trinomial.

3 The quadratic $-5x^2 - 20x - c$ can be written in the form $-5(x + 2)^2$.
What is the value of c?

4 After completing the square $3x^2 - 24x + 57 = 3(x - b)^2 + c$.
What is the value of b and c?

5 The quadratic $2x^2 + kx + 4$ can be written in the form $2\left(x + \frac{5}{2}\right)^2 - \frac{17}{2}$.
What is the value of k?

2. Answers

Greatest Common Factor (GCF)

Time to Practice

1	a	$2(x^2-2)$	b	$3(a^2+3)$	c	$-3(4b^2+1)$
	d	$5(3x^2+5)$	e	$9(3-a^2)$	f	$-16(2b^2+3)$
	g	$25(3x^2-4)$	h	$7(5a^2-7)$	i	$27(3b^2-1)$
2	a	$2x(3x+4)$	b	$2a(2a-1)$	c	$7b(2b-3)$
	d	$12x(1-x)$	e	$3a(3a-5)$	f	$2b(9-2b)$
	g	$2x(x-16)$	h	$3a(4a-9)$	i	$-b(5+2b)$
3	a	$2(x^2+2x+3)$	b	$3(a^2-2a-4)$	c	$12(b^2+2b+3)$
	d	$5(x^2-2x+3)$	e	$6(a^2-3a+4)$	f	$8(b^2+b-4)$
	g	$-4(x^2+2x+3)$	h	$-15(a^2-2a+3)$	i	$-9(b^2-3b+4)$
4	a	$(x+1)(x+9)$	b	$(x+5)(x-2)$	c	$(y-7)(y-4)$
	d	$(y-1)(2y-3)$	e	$(3p+1)(p-8)$	f	$-(2p+3)(7p+5)$
	g	$(4x-9)(2x-3)$	h	$-(3x-5)(4x+1)$		
5	a	$(x+2)(x+5)$	b	$(y-5)(y-9)$	c	$-(x+4)(2x+1)$
	d	$(y-1)(2y-11)$	e	$(4x-1)(4x-3)$	f	$(3y+1)(3y+2)$
	g	$-(2x-3)(5-2x)$ *or* $-(3-2x)(2x-5)$	h	$-(2y+5)(8y+17)$		

Challenge Yourself

1	a	$ab(a-b)$	b	$2xy(xz-2)$	c	$-12mn(m+n)$
	d	$3yz^2(2x-1)$	e	$-2p(pq-2q+3p)$	f	$5ab(a-2b-3)$
	g	$(x-1)(3x+8)$	h	$5(y-2)(3-4y)$ *or* $-5(y-2)(4y-3)$		

Perfect Squares

Time to Practice

1 a $(x+7)^2$ b $(a-3)^2$ c $(b+1)^2$

d $(x-2)^2$ e $-(a+9)^2$ f $(5-b)^2$

g $-(x+8)^2$ h $-(a-11)^2$ i $-(12-b)^2$

2 a $(3x+2)^2$ b $(4a-5)^2$ c $-(5b-3)^2$

d $-(7x+2)^2$ e $(6a+7)^2$ f $-(9b-5)^2$

g $(10x-3)^2$ h $-(2a+5)^2$ i $(1+11b)^2$

3 a $3(x+5)^2$ b $3(a-4)^2$ c $2(b-4)^2$

d $4(x+3)^2$ e $-9(a-3)^2$ f $25(2-b)^2$

g $-16(x+1)^2$ h $10(a-3)^2$ i $-2(5-7b)^2$

Challenge Yourself

1 a $(x^2+5)^2$ b $(2a^4-7)^2$ c $-(4x^3-7)^2$

d $(3x^2-2y^2)^2$ e $(x-1)^2$ f $(a+b+6)^2$

Difference of Two Squares

Time to Practice

1 a $(x-1)(x+1)$ b $(a-3)(a+3)$ c $(b-11)(b+11)$

d $(12-x)(12+x)$ e $(4-a)(4+a)$ f $(1-10b)(1+10b)$

g $(6x-1)(6x+1)$ h $(2a-5)(2a+5)$ i $(9b-4)(9b+4)$

j $(7-3x)(7+3x)$ k $(8-5a)(8+5a)$ l $(6b-1)(6b+1)$

m $12(x-2)(x+2)$ n $-11(a-2)(a+2)$ o $3(b-3)(b+3)$

p $-3(x-5)(x+5)$ q $2(5a-2)(5a+2)$ r $-6(b-3)(b+3)$

s $2(6x-7)(6x+7)$ t $-3(3a-2)(3a+2)$ u $-5(2b-5)(2b+5)$

2 a $x(x+4)$ b $(a-2)(a+12)$ c $(b-3)(b-1)$

d $4x(x-1)$ e $8a(2a-5)$ f $(5b-7)(5b+9)$

g $(5-2x)(2x-1)$ h $(8-7a)(7a+10)$ i $(5-3b)(3b+1)$

j $2(x+2)(x+8)$ k $-3(a-5)(a+1)$ l $-2(6+5b)(8-5b)$

3 a $(x-\sqrt{7})(x+\sqrt{7})$ b $\left(a-\sqrt{3}\right)\left(a+\sqrt{3}\right)$

c $\left(\sqrt{5}-b\right)\left(\sqrt{5+b}\right)$ d $\left(2x-\sqrt{13}\right)\left(2x+\sqrt{13}\right)$

e $\left(3a-\sqrt{19}\right)\left(3a+\sqrt{19}\right)$ f $3\left(b-\sqrt{22}\right)\left(b+\sqrt{22}\right)$

g $-5\left(x-\sqrt{3}\right)\left(x+\sqrt{3}\right)$ h $-2\left(a-\sqrt{2}\right)\left(a+\sqrt{2}\right)$

i $-12\left(2b-\sqrt{3}\right)\left(2b+\sqrt{3}\right)$ j $\left(x+3-\sqrt{6}\right)\left(x+3+\sqrt{6}\right)$

k $\left(\sqrt{15}+4-a\right)\left(\sqrt{15}-4+a\right)$ l $-2(b-13-\sqrt{7})(b-13+\sqrt{7})$

Challenge Yourself

1 a $3a(x-7)(x+7)$ b $4p(m-3n)(m+3n)$ c $(3x-yz)(3x+yz)$

d $\left(2x-\frac{y}{11}\right)\left(2x+\frac{y}{11}\right)$ e $\left(\frac{x}{2}-\frac{3}{5}\right)\left(\frac{x}{2}+\frac{3}{5}\right)$ f $\frac{5}{2}(x-1)(x+1)$

g $\left(\frac{ab}{6}-\frac{c}{4}\right)\left(\frac{ab}{6}+\frac{c}{4}\right)$ h $\left(\frac{pq}{9}-\frac{mn}{10}\right)\left(\frac{pq}{9}+\frac{mn}{10}\right)$

The AC Method

Time to Practice

1	a	$(x+1)(x+3)$	b	$(a-5)(a-2)$	c	$(b-6)(b+2)$
	d	$(x-5)(x-3)$	e	$(a-5)(a-4)$	f	$(b-10)(b+11)$
	g	$-(x-4)(x-2)$	h	$-(a-2)(a+5)$	i	$-(b-3)(b+8)$
	j	$4(x+1)(x+4)$	k	$6(a-3)(a+1)$	l	$2(b-7)(b-6)$
2	a	$6(x+1)(x+2)$	b	$-3(a-1)(a+6)$	c	$-2(b-4)(b-3)$
	d	$9(x-2)(x-1)$	e	$-5(a+2)(a+3)$	f	$-4(b-4)(b-3)$
	g	$15(x+1)(x+2)$	h	$-9(a-2)(a-1)$	i	$-6(b-7)(b-1)$
3	a	$(x+1)(2x+3)$	b	$(a-2)(3a-2)$	c	$(2b-1)(2b+3)$
	d	$(3x+1)(2x+5)$	e	$(2a-1)(3a+5)$	f	$(4b+1)(2b+3)$
	g	$(3x-1)(4x+3)$	h	$-(3a-2)(3a+1)$	i	$-(2b-5)(3b+1)$
	j	$-(x-2)(3x-5)$	k	$-(2a+3)(a+6)$	l	$-(b-2)(4b+1)$
4	a	$2(x-2)(3x+1)$	b	$-3(2a-1)(a+2)$	c	$4(2b-3)(b+1)$
	d	$-5(3x-1)(x+2)$	e	$7(2a+1)(a+3)$	f	$-6(2b-5)(b-2)$
	g	$11(2x-7)(2x+1)$	h	$5(3a-2)(2a+3)$	i	$-2(4b-3)(2b-1)$
5	a	$6(x-1)(3x+5)$	b	$-2(a-2)(3a+5)$	c	$-4(2b-3)(b+4)$
	d	$7(x-3)(3x+1)$	e	$-6(2a-1)(a+5)$	f	$-2(b-4)(3b+4)$
	g	$9(2x-1)(x+4)$	h	$-9(3b-1)(b+2)$	i	$-8(4a-1)(3a+1)$

Challenge Yourself

1	$k=-5$	2	$k=6$	3	$m=1, k=4$
4	$p=3, k=-13$	5	$k=8 \text{ or } 16$		

Completing The Square

Time to Practice

1 a $x(x+12)$ b $a(a-10)$ c $b(b-2)$

d $x(x+8)$ e $a(a-4)$ f $b(b-16)$

g $x(x+9)$ h $a(a-7)$ i $b(b+11)$

j $x(x-1)$ k $a(a-3)$ l $b(b-5)$

2 a $\left(x+2-\sqrt{3}\right)\left(x+2+\sqrt{3}\right)$ b $(a+3-\sqrt{7})(a+3+\sqrt{7})$

c $\left(b+1-\sqrt{5}\right)\left(b+1+\sqrt{5}\right)$ d $\left(x+5-\sqrt{29}\right)\left(x+5+\sqrt{29}\right)$

e $\left(a-4-\sqrt{3}\right)\left(a-4+\sqrt{3}\right)$ f $\left(b-6-\sqrt{26}\right)\left(b-6+\sqrt{26}\right)$

g $(x-2-\sqrt{7})(x-2+\sqrt{7})$ h $\left(a-4-\sqrt{21}\right)\left(a-4+\sqrt{21}\right)$

i $\left(b+7-\sqrt{43}\right)\left(b+7+\sqrt{43}\right)$ j $-\left(x-2-\sqrt{11}\right)\left(x-2+\sqrt{11}\right)$

k $-\left(a+1-\sqrt{3}\right)\left(a+1+\sqrt{3}\right)$ l $-\left(b-5-\sqrt{13}\right)\left(b-5+\sqrt{13}\right)$

3 a $2\left(x+1-\sqrt{3}\right)\left(x+1+\sqrt{3}\right)$ b $4\left(a-1-\sqrt{6}\right)\left(a-1+\sqrt{6}\right)$

c $5\left(b+2-\sqrt{5}\right)\left(b+2+\sqrt{5}\right)$ d $3\left(x-2-\sqrt{17}\right)\left(x-2+\sqrt{17}\right)$

e $5(a-3-\sqrt{7})(a-3+\sqrt{7})$ f $2\left(b-7-\sqrt{2}\right)\left(b-7+\sqrt{2}\right)$

g $-3\left(x+5-\sqrt{19}\right)\left(x+5+\sqrt{19}\right)$ h $-2\left(a-2-\sqrt{11}\right)\left(a-2+\sqrt{11}\right)$

i $-4\left(b-1-\sqrt{5}\right)\left(b-1+\sqrt{5}\right)$

4 a $3\left(x-\frac{2-\sqrt{7}}{3}\right)\left(x-\frac{2+\sqrt{7}}{3}\right)$ b $5\left(a-\frac{1-\sqrt{6}}{5}\right)\left(a-\frac{1+\sqrt{6}}{5}\right)$

c $8\left(b-\frac{3-\sqrt{11}}{4}\right)\left(b-\frac{3+\sqrt{11}}{4}\right)$ d $-2\left(x+\frac{5-\sqrt{3}}{2}\right)\left(x+\frac{5+\sqrt{3}}{2}\right)$

e $3\left(a-\frac{1-\sqrt{7}}{3}\right)\left(a-\frac{1+\sqrt{7}}{3}\right)$ f $-3\left(b-\frac{1-\sqrt{7}}{3}\right)\left(b-\frac{1+\sqrt{7}}{3}\right)$

Completing The Square

4 g $-3\left(x+\frac{1-\sqrt{10}}{3}\right)\left(x+\frac{1+\sqrt{10}}{3}\right)$ h $-9\left(a-\frac{3-\sqrt{5}}{3}\right)\left(a-\frac{3+\sqrt{5}}{3}\right)$

i $4\left(b+\frac{4-\sqrt{13}}{2}\right)\left(b+\frac{4+\sqrt{13}}{2}\right)$

Challenge Yourself

1 a $\left(x-\frac{4-\sqrt{10}}{2}\right)\left(x-\frac{4+\sqrt{10}}{2}\right)$ b $-\left(x-\frac{9-\sqrt{33}}{3}\right)\left(x-\frac{9+\sqrt{33}}{3}\right)$

c $-3\left(x+\frac{4-\sqrt{42}}{6}\right)\left(x+\frac{4+\sqrt{42}}{6}\right)$ d $2\left(a+\frac{3-2\sqrt{21}}{6}\right)\left(a+\frac{3+2\sqrt{21}}{6}\right)$

e $3\left(a-\frac{5-3\sqrt{5}}{15}\right)\left(a-\frac{5+3\sqrt{5}}{15}\right)$ f $-6\left(x-\frac{1-\sqrt{3}}{6}\right)\left(x-\frac{1+\sqrt{3}}{6}\right)$

2 $c = 16$ 3 $c = 20$

4 $b = 4 \text{ and } c = 9$ 5 $k = 10$

Solving
Quadratics

Solving Quadratics: 0, 1, or 2 Solutions (No More, No Less)

You're now good at factoring.

Next is solving quadratic equations.

Everything from factoring gets used here.

Plus an extra bit to solve it.

The only exception is using the quadratic formula.

For that.

No need to factor.

You just plug numbers into the formula and simplify.

Depending on how the quadratic is written.

A different way is used to solve it.

Here are the 5 ways.

1) Square root method
2) Completing the square
3) Null Factor Law
4) Factoring
5) Quadratic formula

We'll go through them one by one.

By the end.

As soon as you see a quadratic.

You'll know exactly how to solve it.

Before getting into it.

This part's important.

Quadratics aren't like linear equations where there's always one solution.

With quadratics there can be.

No solution.

One solution.

Or two solutions.

Meaning.

There's no number that makes the equation true.

Only one number works.

Or there are two numbers that work.

Square Root Method: What Will It Be, Positive, Negative or Both?

The square root method is used to solve a quadratic,

when a squared term equals a number.

Or when it's a difference of two squares.

Here's what to do.

Get the squared term by itself on one side.

And the numbers on the other.

From there, take the square root of both sides.

Taking the square root is the opposite of squaring.

What this does is it cancels out the power of 2 from the squared term.

So you're just left with what's being squared.

However.

On the number side.

When taking the square root.

Write it as both the positive OR negative square root of that number.

The plus or minus goes in front of the root sign.

You have to include both.

Otherwise, the answer won't be complete.

From there simplify it if you can.

Keeping the $\pm$ in front.

Opposite Operations

$+$	$\longleftrightarrow$	$-$
$\times$	$\longleftrightarrow$	$\div$
$(\ )^2$	$\longleftrightarrow$	$\sqrt{\ }$

If the squared term is just a letter squared.

You're done.

But if the squared term is something inside brackets.

You still have to solve for the variable.

To solve.

Write two equations.

One where what's inside the brackets equals the positive number.

And one where it equals the negative number.

Then solve each equation separately.

Why do you have to write both plus and minus?

When solving an equation.

You always do this.

Because you've got something squared equals a number.

That "something" could be positive.

Or it could be negative.

A positive number times itself is positive.

A negative number times itself is also positive.

So to cover both cases.

We write the plus and the minus sign.

Except for zero.

The square root of zero is just zero.

No need to write the plus and minus.

One final point.

If the number side is negative.

You can't take the square root of that number.

The square root of a negative number doesn't exist.

No number multiplied by itself will equal a negative.

So there's no solution.

In this case, just write 'no solution'.

That's the answer.

Later, in advanced maths.

There's this thing called *imaginary* numbers.

It's basically a workaround for square roots of negative numbers.

But for now.

Just write 'no solution'.

Alright, here are the steps.

1) Get the squared term by itself on one side
2) Take the square root of both sides.
3) Cancel out the power of 2 and write other side as the $\pm\sqrt{number}$
4) Simplify the square root (if possible)
5) If the squared term is an expression write two separate equations. One equals the positive root, the other the negative root.
6) Solve each equation

Worked Example 1

Solve the following quadratic equations.

a $x^2 - 9 = 0$ b $2x^2 - 14 = 0$ c $x^2 + 25 = 0$

Explanation		Solution
a	Write the equation.	$x^2 - 9 = 0$
	Move the constant to the right.	$x^2 = 9$
	Take the square root of both sides.	$x = \pm\sqrt{9}$
	Simplify.	$x = \pm 3 \quad (x = 3 \text{ or } x = -3)$
b	Write the equation.	$2x^2 - 14 = 0$
	Move the constant to the right.	$2x^2 = 14$
	Move the coefficient to the right.	$x^2 = 7$
	Take the square root of both sides.	$x = \pm\sqrt{7} \quad (x = \sqrt{7} \text{ or } x = -\sqrt{7})$
c	Write the equation.	$x^2 + 25 = 0$
	Move constant to the right.	$x^2 = -25$
		$x = \sqrt{-25}$
	We can't take the square root of a negative number.	*No Solution*

Worked Example 2

Solve the following quadratic equations.

a $\mathbf{49x^2 = 1}$

b $\mathbf{75 - 12x^2 = 0}$

Explanation	**Solution**
a Write the equation.	$49x^2 = 1$
Move the coefficient to the right.	$x^2 = \frac{1}{49}$
Take the square root of both sides.	$x = \pm\sqrt{\frac{1}{49}}$
Simplify.	$x = \pm\frac{1}{7} \quad \left(x = \frac{1}{7} \text{ or } x = -\frac{1}{7}\right)$
b Write the equation.	$75 - 12x^2 = 0$
Move variable to the right.	$75 = 12x^2$
Move the coefficient to the other side.	$\frac{75}{12} = x^2$
Simplify fraction.	$x^2 = \frac{25}{4}$
Take the square root of both sides.	$x = \pm\sqrt{\frac{25}{4}}$
Simplify the square root.	$x = \pm\frac{5}{2} \quad \left(x = \frac{5}{2} \text{ or } x = -\frac{5}{2}\right)$

Worked Example 3

Solve the following quadratic equations.

a $(x-2)^2 = 0$ b $(5-x)^2 = 0$ b $(3x-7)^2 = 0$

Explanation		Solution
a	Write the equation.	$(x-2)^2 = 0$
	Take the square root of both sides.	$x - 2 = 0$
	Move the constant to the right.	$x = 2$
b	Write the equation.	$(5-x)^2 = 0$
	Take the square root of both sides.	$5 - x = 0$
	Move variable to the right.	$x = 5$
c	Write the equation.	$(3x-7)^2 = 0$
	Take the square root of both sides.	$3x - 7 = 0$
	Move the constant to the right.	$3x = 7$
	Move the coefficient to the right.	$x = \frac{7}{3}$

Writing the square root over the numerator only and not covering the entire fraction. $\sqrt{\frac{a}{b}} \neq \frac{\sqrt{a}}{b}$

Worked Example 4

Solve the following quadratic equations.

a $(x-3)^2 - 9 = 0$

b $2(x+1)^2 - 8 = 0$

Explanation		Solution
a	Write the equation.	$(x-3)^2 - 9 = 0$
	Move the constant to the right.	$(x-3)^2 = 9$
	Take the square root of both sides.	$x - 3 = \pm 3$
	Rewrite as two separate equations and solve each for x.	$x - 3 = -3$ *or* $x - 3 = 3$ $x = 0$ $\quad$ $x = 6$
	Write answer.	$x = 0$ *or* $x = 6$
b	Write the equation.	$2(x+1)^2 - 8 = 0$
	Move the constant to the right.	$2(x+1)^2 = 8$
	Move the number outside the brackets to the other side.	$(x+1)^2 = 4$
	Take the square root of both sides.	$x + 1 = \pm 2$
	Rewrite as two separate equations and solve each for x.	$x + 1 = -2$ *or* $x + 1 = 2$ $x = -3$ $\quad$ $x = 1$
	Write answer.	$x = -3$ *or* $x = 1$

Worked Example 5

Solve the following quadratic equations.

a $(3x-1)^2-4=0$

b $\frac{1}{4}(5x-2)^2-4=0$

Explanation	Solution
a Write the equation.	$(3x-1)^2-4=0$
Move the constant to the right.	$(3x-1)^2=4$
Take the square root of both sides.	$3x-1=\pm 2$
Rewrite as two separate equations and solve each for x.	$3x-1=-2$ *or* $3x-1=2$ $3x=-1$ $3x=3$ $x=-\frac{1}{3}$ $x=1$
Write answer.	$x=-\frac{1}{3}$ *or* $x=1$
b Write the equation.	$\frac{1}{4}(5x-2)^2-4=0$
Move the constant to the right.	$\frac{1}{4}(5x-2)^2=4$
Move the denominator to the right.	$(5x-2)^2=16$
Take the square root of both sides.	$5x-2=\pm 4$
Rewrite as two separate equations and solve each for x.	$5x-2=-4$ *or* $5x-2=4$ $5x=-2$ $5x=6$ $x=-\frac{2}{5}$ $x=\frac{6}{5}$
Write answer.	$x=-\frac{2}{5}$ *or* $x=\frac{6}{5}$

Worked Example 6

Solve the following quadratic equation: $45 - \left(x - \frac{7}{2}\right)^2 = 0$

Explanation	Solution
Write the equation.	$45 - \left(x - \frac{7}{2}\right)^2 = 0$
Move brackets to the right.	$45 = \left(x - \frac{7}{2}\right)^2$
Take the square root of both sides.	$x - \frac{7}{2} = \pm\sqrt{45}$
(Simplify the surd)	$\sqrt{45} = \sqrt{9 \times 5} = 3\sqrt{5}$
Rewrite equation in simplified surd form.	$x - \frac{7}{2} = \pm 3\sqrt{5}$
Move constant to the right.	$x = \frac{7}{2} \pm 3\sqrt{5}$
Write the surd as a fraction.	$x = \frac{7}{2} \pm \frac{3\sqrt{5}}{1}$
Combine fractions.	$x = \frac{7 \pm 6\sqrt{5}}{2}$

If the square root doesn't simplify to a whole number, just solve for the variable and keep the ± sign in the answer. Don't split it into two equations.

Your Turn To Try

1 **Which equation is best solved using the square root method?**

A $a^2 + 5a - 6$

B $x^2 - 64$

C $(a-2)(a-5)$

D $x^2 - 64x$

2 **The solution(s) to the equation $2x^2 - 18 = 0$ are?**

A $x = \pm 9$

B $x = 9$

C $x = \pm 3$

D $x = 3$

3 **For the equation $4x^2 - 1 = 63$, what are the value(s) of x?**

A $x = -4$

B $x = -4 \; or \; x = 4$

C $x = 4$

D $x = -16 \; or \; x = 16$

4 **What's the solution to $x^2 + 25 = 0$?**

A $x = 5$

B $x = \pm 5$

C $x = -5$

D $No \; Solution$

5 **What are the solutions to $2(x+1)^2 = 32$?**

A $x = -5 \; or \; x = 3$

B $x = -3 \; or \; x = 5$

C $x = \pm 3$

D $x = -2 \; or \; x = 4$

Question	1	2	3	4	5
Answer	B	C	B	D	A

Time To Practice

1 Solve the following quadratic equations for x.

a $x^2 - 16 = 0$ b $a^2 - 64 = 0$ c $b^2 - 20 = 5$

d $x^2 - 121 = 0$ e $6a^2 - 54 = 0$ f $16b^2 - 16 = 0$

g $200 - 2x^2 = 0$ h $72 - 2a^2 = 0$ i $-7b^2 + 49 = 0$

j $x^2 + 9 = 0$ k $a^2 + 4 = 0$ l $8b^2 + 72 = 0$

m $x^2 - 10 = 0$ n $3a^2 - 45 = 0$ o $22 - 2b^2 = 0$

2 Solve the following for x.

a $x^2 = \frac{1}{4}$ b $4a^2 = \frac{9}{16}$ c $4b^2 - \frac{1}{9} = 0$

d $9x^2 - 4 = 0$ e $36a^2 - 49 = 0$ f $100 - 81b^2 = 0$

3 Solve the following for x.

a $(x + 4)^2 = 0$ b $(a - 5)^2 = 0$ c $(2 - b)^2 = 0$

d $(2x + 3)^2 = 0$ e $(7 - 6a)^2 = 0$ f $(-4b - 1)^2 = 0$

g $(x + 3)^2 - 9 = 0$ h $(a - 5)^2 - 16 = 0$ i $(7 - b)^2 - 36 = 0$

4 Solve the following for x.

a $2(x + 1)^2 - 8 = 0$ b $5(a + 3)^2 - 20 = 0$ c $3(b - 5)^2 - 27 = 0$

d $-3(x - 4)^2 + 12 = 0$ e $4(2 - a)^2 - 16 = 0$ f $-(3b - 4)^2 + 9 = 0$

g $25 - (2x + 1)^2 = 0$ h $6(7 - 2a)^2 = 24$ i $-3(4b - 5)^2 = -75$

j $2(3x - 5)^2 - 50 = 0$ k $-5(1 - 2a)^2 = -80$ l $-6(5b - 1)^2 = -54$

Challenge Yourself

1 Solve each of the following quadratic equations.

a $-5(1-2x)^2 = -40$

b $-2(2m-1)^2 = -36$

c $\frac{1}{3}(2x-1)^2 - 5 = 0$

d $-\frac{3}{4}(1-2n)^2 + 9 = 0$

e $\frac{5}{3}(3p+2)^2 - 40 = 0$

f $-\frac{1}{4}(3x+4)^2 + 7 = 0$

g $\frac{3}{2}\left(x-\frac{1}{3}\right)^2 - 27 = 0$

h $30 - \frac{2}{3}\left(q+\frac{1}{2}\right)^2 = 0$

2 The simplified form of $(2x-1)^2 - (x-k)(x+k) = 3x^2 - 4x + 10$.
Find k where k is a constant.

Completing the Square: A Sweet Mix of Factoring and Solving

When solving quadratics by completing the square.

Rewrite it by completing the square.

Get the quadratic in the form of something squared equals a number.

Or into a difference of two squares form.

Don't fully factor it into two brackets.

From there.

Solve using the square root method.

That's it.

But only solve this way if the question asks you to use Completing the Square.

Otherwise, use the quadratic formula.

If the quadratic can be factored.

Factor it first.

Using the other factoring techniques.

Then use the Null Factor Law to solve.

Which we'll look into next.

Worked Example 1

Use completing the square to find exact solutions for $x^2 + 8x + 1 = 0$

Explanation	Solution
Write the equation.	$x^2 + 8x + 1 = 0$
Half the b number and square it.	$\rightarrow \left(\frac{8}{2}\right)^2 \rightarrow (4)^2$
Add and then subtract this squared number after the middle term.	$x^2 + 8x + (4)^2 - (4)^2 + 1 = 0$
Group the first three terms inside brackets.	$[x^2 + 8x + (4)^2] - 16 + 1 = 0$
Factor the perfect square part and combine the leftover numbers.	$(x + 4)^2 - 15 = 0$
Move the constant to the right side.	$(x + 4)^2 = 15$
Take the square root of both sides.	$x + 4 = \pm\sqrt{15}$
Move the constant to the right side.	$x = -4 \pm \sqrt{15}$

Exact solutions means the answer needs to be a whole number or in surd form.

Worked Example 2

Use completing the square to find exact solutions for $x^2 - 6x = 2$

Explanation	**Solution**
Write the equation.	$x^2 - 6x = 2$
Half the b number and square it.	$\rightarrow \left(\frac{6}{2}\right)^2 \rightarrow (3)^2$
Add this squared number after the middle term and on the other side.	$x^2 - 6x + (3)^2 = 2 + (3)^2$
Group the left three terms inside brackets.	$[x^2 - 6x + (3)^2] = 2 + 9$
Factor the perfect square part and combine the numbers.	$(x - 3)^2 = 11$
Take the square root of both sides.	$x - 3 = \pm\sqrt{11}$
Move the constant to the right side.	$x = 3 \pm \sqrt{11}$

When completing the square with equations, you don't have to add and then subtract $\left(\frac{b}{2}\right)^2$. You can just add it to both side.

Worked Example 3

Solve the equation $x^2 - 4 = 2x$ by completing the square. Give exact answers.

Explanation	Solution
Rewrite the equation in standard form.	$x^2 - 2x - 4 = 0$
Half the b number and square it.	$\rightarrow \left(\frac{2}{2}\right)^2 \rightarrow (1)^2$
Add and then subtract this squared number after the middle term.	$x^2 - 2x + (1)^2 - (1)^2 - 4 = 0$
Group the first three terms inside brackets.	$[x^2 - 2x + (1)^2] - 1 - 4 = 0$
Factor the perfect square part and combine the leftover numbers.	$(x - 1)^2 - 5 = 0$
Move the constant to the right side.	$(x - 1)^2 = 5$
Take the square root of both sides.	$x - 1 = \pm\sqrt{5}$
Move the constant to the right side.	$x = 1 \pm \sqrt{5}$

If the GCF is a number, divide everything by it. Zero divided by any number is still zero, so the zero on the other side won't change.

Worked Example 4

Use completing the square to find exact solutions for $2x^2 - 12x + 17 = 0$

Explanation	**Solution**
Write the equation.	$2x^2 - 12x + 17 = 0$
Factor out a.	$2\left(x^2 - 6x + \frac{17}{2}\right) = 0$
Move number in front of brackets to right side.	$x^2 - 6x + \frac{17}{2} = 0$
Add and then subtract $\left(\frac{b}{2}\right)^2$ after the middle term.	$x^2 - 6x + (3)^2 - (3)^2 + \frac{17}{2} = 0$
Group the first three terms inside brackets.	$[x^2 - 6x + (3)^2] - \frac{18}{2} + \frac{17}{2} = 0$
Factor the perfect square part and combine the leftover numbers.	$[x^2 - 6x + (3)^2] - \frac{1}{2} = 0$
Move the constant to the right.	$(x-3)^2 - \frac{1}{2} = 0$
Take the square root of both sides.	$(x-3)^2 = \frac{1}{2}$
Rationalize the denominator.	$x - 3 = \pm\frac{1}{\sqrt{2}}$
Rewrite solutions in rationalized form.	$\frac{1}{\sqrt{2}} = \frac{1}{\sqrt{2}} \times \frac{\sqrt{2}}{\sqrt{2}} = \frac{\sqrt{2}}{2}$
Move the constant to the right.	$x - 3 = \pm\frac{\sqrt{2}}{2}$
Combine as one fraction. (optional)	$x = 3 \pm \frac{\sqrt{2}}{2}$
	$x = \frac{6 \pm \sqrt{2}}{2}$

Worked Example 5

Use completing the square to find exact solutions for $3x^2 = 7x + 3$

Explanation	**Solution**
Write in standard form and make one side equal zero.	$3x^2 - 7x - 3 = 0$
Factor out a.	$3\left(x^2 - \frac{7}{3}x - 1\right) = 0$
Move number in front of brackets to right side.	$x^2 - \frac{7}{3}x - 1 = 0$
Add and then subtract $\left(\frac{b}{2}\right)^2$ after the middle term.	$\left(x^2 - \frac{7}{3}x + \left(\frac{7}{6}\right)^2 - \left(\frac{7}{6}\right)^2 - 1\right) = 0$
Group the first three terms inside brackets.	$\left(\left[x^2 - \frac{7}{3}x + \left(\frac{7}{6}\right)^2\right] - \frac{49}{36} - \frac{36}{36}\right) = 0$
Factor the perfect square part and combine the leftover numbers.	$\left(x - \frac{7}{6}\right)^2 - \frac{85}{36} = 0$
Move the constant to the right.	$\left(x - \frac{7}{6}\right)^2 = \frac{85}{36}$
Take the square root of both sides.	$x - \frac{7}{6} = \pm\frac{\sqrt{85}}{6}$
Move the constant to the right.	$x = \frac{7}{6} \pm \frac{\sqrt{85}}{6}$
Combine as one fraction. (optional)	$x = \frac{7 \pm \sqrt{85}}{6}$

Worked Example 6

If completing the square on $2x^2 - bx + 29 = 0$ results in $2(x-4)^2 = 3$, what is the value of b?

Explanation	Solution
Write the second equation.	$2(x-4)^2 = 3$
Expand the brackets.	$2(x^2 - 8x + 16) = 3$
Multiply the number outside by each term inside.	$2x^2 - 16x + 32 = 3$
Move the number on the right to the left.	$2x^2 - 16x + 32 - 3 = 0$
Combine like terms.	$2x^2 - 16x + 29 = 0$
Let the two expressions equal.	$2x^2 - 16x + 29 = 2x^2 - bx + 29$
Match the coefficients.	$\therefore b = 16$

When two expressions are written in the same form, you can compare the coefficients of each matching term. That's is called *equating the coefficients*.

Your Turn To Try

1 **Given the equation $x^2 + 8x + 5 = 0$. Which is the correct completed square form?**

A $(x+4)^2 + 11 = 0$ B $(x-4)^2 - 11 = 0$

C $(x+4)^2 - 11 = 0$ D $(x-4)^2 + 11 = 0$

2 **Which equation is equivalent to $x^2 + 10x + 21 = 0$?**

A $(x+5)^2 - 4 = 0$ B $(x+5)^2 + 4 = 0$

C $(x-5)^2 - 4 = 0$ D $(x-5)^2 - 4 = 0$

3 **After completing the square, $3(x-1)^2 = 27$. What are the solutions?**

A $x = 2 \text{ or } x = -4$ B $x = -2 \text{ or } x = 4$

C $x = -2 \text{ or } x = -4$ D $x = 2 \text{ or } x = 4$

4 **The solutions to $x^2 = 4x + 1$ when when solved by completing the square are?**

A $x = 2 - \sqrt{5}$ B $x = -2 \pm \sqrt{5}$

C $x = 2 + \sqrt{5}$ D $x = 2 \pm \sqrt{5}$

5 **Using complete the square what are the solutions to $x^2 - 8x - 7 = 0$?**

A $x = 4 \pm \sqrt{23}$ B $x = 4 + \sqrt{23}$

C $x = -4 \pm \sqrt{23}$ D $x = 4 - \sqrt{23}$

Question	1	2	3	4	5
Answer	C	A	B	D	A

Time To Practice

1 Solve the following equations by completing the square. Give exact answers.

a $x^2 + 2x - 2 = 0$ b $a^2 - 2a - 5 = 0$ c $b^2 + 2b - 1 = 0$

d $x^2 - 2x - 4 = 0$ e $a^2 + 6a + 7 = 0$ f $b^2 - 8b + 9 = 0$

g $10 - 6x - x^2 = 0$ h $10a + 4 - a^2 = 0$ i $-1 - 12b - b^2 = 0$

2 Solve the following equations by completing the square. Give exact answers.

a $x^2 - x - 1 = 0$ b $a^2 + 3a - 2 = 0$ c $b^2 - 5b - 3 = 0$

d $x^2 - 3x + 1 = 0$ e $a^2 + 5a + 3 = 0$ f $b^2 + 7b - 1 = 0$

g $x^2 + 3x - 6 = 0$ h $a^2 - 9a + 7 = 0$ i $-b^2 = 20 - 11b$

j $5x - x^2 = -10$ k $11a - 25 = a^2$ l $7b + 9 = -b^2$

3 Solve the following equations by completing the square. Give exact answers.

a $2x^2 + 4x - 8 = 0$ b $3a^2 + 12a - 6 = 0$ c $4b^2 - 8b - 8 = 0$

d $5x^2 - 15x - 15 = 0$ e $2a^2 - 6a + 2 = 0$ f $3b^2 - 9b - 3 = 0$

g $-5x^2 - 15x = -25$ h $21 - 7a = 7a^2$ i $-2b^2 - 18b = 38$

4 For the following, find exact solutions (if they exist) by completing the square.

a $5x^2 - 2x - 4 = 0$ b $3a^2 + 2a - 4 = 0$ c $4b^2 - 2b = 5$

d $5x = 3x^2 - 3$ e $4a - 3a^2 + 2 = 0$ f $2b - 5b^2 = -1$

g $-2 = 7x - 3x^2$ h $4a^2 - 4a = -7$ i $-3 - 5b^2 = -b$

Challenge Yourself

1 Solve each of the following equations for x using complete the square. Give exact answers.

a $x^2 + 4x = \frac{29}{4}$

b $a^2 + 6a = -\frac{53}{9}$

c $x^2 - 5x + \frac{1}{16} = 0$

d $3a^2 + 4a + \frac{10}{9} = 0$

2 Write $x^2 + 6x - 2 = 0$ in the form $(x + p)^2 + q = 0$.
What is the value of p and q.

3 Given that completing the square for $3x^2 + bx - 5 = 0$ results in $3(x + 2)^2 = 17$.
What is the value of b.

4 For the quadratic $x^2 + x + m = 0$ completing the square leads to $\left(x + \frac{1}{2}\right)^2 = 4$. What is the value of m.

5 Write $-3x^2 - 6x + 4 = 0$ in the form $a(x + b)^2 + c = 0$.
What are the values of a, b and c.

6 Write $2x^2 - 12x + 21 = 0$ in the form $a(x + b)^2 + c = 0$.
What are the values of a, b and c.

Null Factor Law: One's Gotta Be Zero. No Ifs, Ands, or Buts

When a quadratic can be written in factored form.

The Null Factor Law is used to solve it.

Basically, this law says.

If two numbers multiply to equal zero.

Then one of those numbers must be zero.

For quadratics, here's how it works.

First.

The quadratic is set to equal zero.

Once it's factored.

One of those factors must be zero.

Either the first one.

Or the second one.

Each factor is then set to equal zero.

And the variable is solved for each one.

Why zero?

Because zero is a special number.

Anything times zero is zero.

It's the only number that works like that.

And that makes it easy to work with.

Think about it.

Let's say two numbers multiply to 1.

We wouldn't know for sure what they are.

Could be 1×1.

or -1×-1.

Which is it?

Can't say for sure.

We'd be guessing.

Pick any other number.

What about 6?

You can get it from 2×3.

-2×-3.

1×6.

-1×-6.

And that's not even including fractions or surds.

Again, which is it?

Can't say for sure.

We'd be guessing.

Zero is different.

It's the only number we know for sure.

When two numbers multiply to zero.

One of them must be zero.

No guessing at all.

We'll start with quadratics that are already factored.

Just to get used to how the Null Factor Law works.

Then we'll move on to the ones you have to factor first.

Here are the steps.

1) Make the equation equal 0 (if it isn't already)
2) Set each factor equal to 0
3) Solve each factor for the variable
4) Write both solutions with 'or' between them

Worked Example 1

Solve these quadratic equations using the Null Factor Law.

a $x(x+2)=0$ b $x(x-5)=0$ c $-3x(x-1)=0$

Explanation	**Solution**		
a Write the equation	$x(x+2)=0$		
Solve each factor for x.	$x=0$	or	$x+2=0$
			$x=-2$
Write answer.	$x=0 \ or \ x=-2$		
b Write the equation	$x(x-5)=0$		
Solve each factor for x.	$x=0$	or	$x-5=0$
			$x=5$
Write answer.	$x=0 \ or \ x=5$		
c Write the equation	$-3x(x-1)=0$		
Solve each factor for x.	$-3x=0$	or	$x-1=0$
	$x=0$		$x=1$
Write answer.	$x=0 \ or \ x=1$		

Writing "and" instead of "or" between solutions. ⚠

Worked Example 2

Solve these quadratic equations using the Null Factor Law.

a $(x-1)(x+1)=0$ b $(x-6)(x+2)=0$ c $(7+x)(23-x)=0$

Explanation	Solution		
a Write the equation	$(x-1)(x+1)=0$		
Solve each factor for x.	$x-1=0$	or	$x+1=0$
	$x=1$		$x=-1$
Write answer.	$x=1$ or $x=-1$		
b Write the equation	$(x-6)(x+2)=0$		
Solve each factor for x.	$x-6=0$	or	$x+2=0$
	$x=6$		$x=-2$
Write answer.	$x=6$ or $x=-2$		
c Write the equation	$(7+x)(23-x)=0$		
Solve each factor for x.	$7+x=0$	or	$23-x=0$
	$x=-7$		$x=23$
Write answer.	$x=-7$ or $x=23$		

Worked Example 3

Solve the following quadratic equations.

a $-(4x-3)(3x-5)=0$

b $-2(7x+12)(9x+2)=0$

Explanation	Solution
a Write the equation	$-(4x-3)(3x-5)=0$
Solve each factor for x.	$4x-3=0$ or $3x-5=0$
	$4x=3$ $3x=5$
	$x=\frac{3}{4}$ $x=\frac{5}{3}$
Write answer.	$x=\frac{3}{4}\ or\ x=\frac{5}{3}$
b Write the equation	$-2(7x+12)(9x+2)=0$
Solve each factor for x.	$7x+12=0$ or $9x+2=0$
	$7x=-12$ $9x=-2$
	$x=-\frac{12}{7}$ $x=-\frac{2}{9}$
Write answer.	$x=-\frac{12}{7}\ or\ x=-\frac{2}{9}$

Worked Example 4

Given the quadratic equation has solutions $x = -\frac{1}{3}$, $x = \frac{5}{2}$, and the equation is $ax^2 + bx + c = 0$. Find the value of a, b and c?

Explanation	Solution
Write each solution.	$x = -\frac{1}{3}$ or $x = \frac{5}{2}$
Move the denominator to the left side.	$3x = -1$ $2x = 5$
Move the number to the left.	$3x + 1 = 0$ $2x - 5 = 0$
Write quadratic in factored form.	$(3x + 1)(2x - 5) = 0$
Expand brackets.	$6x^2 - 15x + 2x - 5 = 0$
Collect like terms.	$6x^2 - 13x - 5 = 0$
Let the two expressions equal.	$6x^2 - 13x - 5 = ax^2 + bx + c$
Match the coefficients.	$\therefore a = 6, b = -13, c = -5$

If you rearrange a solution, like $x = something$, so it's zero on one side.
The expression side is a factor.

Your Turn To Try

1 **Which solution satisfies the equation $(x+2)(x-3)=0$?**

A $x=-3$
B $x=-2$
C $x=2$
D $x=-1$

2 **What are the solutions to $x(x+7)=0$?**

A $x=-7$
B $x=0 \text{ or } x=7$
C $x=0$
D $x=0 \text{ or } x=-7$

3 **The solutions to $(3m-6)(m+3)=0$ are?**

A $m=-2 \text{ or } m=-3$
B $m=6 \text{ or } m=-3$
C $m=2 \text{ or } m=-3$
D $m=2 \text{ or } m=3$

4 **Which equation has solutions $x=4 \text{ or } x=-2$?**

A $(x-4)(x+2)=0$
B $(x+4)(x+2)=0$
C $(x-4)(x-2)=0$
D $(x+4)(x-2)=0$

5 **Which factored form would give solutions of $x=\frac{1}{3} \text{ or } x=2$?**

A $(3x+1)(x-2)=0$
B $(x-3)(x-2)=0$
C $(3x-1)(x-2)=0$
D $(3x-1)(x+2)=0$

Question	1	2	3	4	5
Answer	B	D	C	A	C

Time To Practice

1 Solve the following quadratic equations using the Null Factor Law.

a $x(x+3)=0$

b $a(a+5)=0$

c $b(b-7)=0$

d $3x(x+2)=0$

e $4a(a-3)=0$

f $-b(b-9)=0$

g $-6x(4-x)=0$

h $-8a(1-a)=0$

i $-3b(-8-b)=0$

2 Solve the following quadratic equations using the Null Factor Law.

a $(x-5)(x+3)=0$

b $(a-4)(a+1)=0$

c $(b-2)(b-6)=0$

d $(4-x)(x-8)=0$

e $(a-8)(3-a)=0$

f $(b-3)(b-9)=0$

g $(x-4)(x+3)=0$

h $(2-a)(9-a)=0$

i $(5-b)(6-b)=0$

3 Solve the following quadratic equations.

a $2(3x+1)(x-4)=0$

b $-3(2m-5)(m+6)=0$

c $4(p-2)(5p-3)=0$

d $-2(4a+1)(2a-7)=0$

e $3(2u+5)(u-1)=0$

f $-5(w-3)(3w+4)=0$

g $-7(5-3x)(3x+2)=0$

h $-6(7-2y)(4-3y)=0$

i $-9(4w+3)(2w-9)=0$

j $8(1-7t)(2t-11)=0$

Challenge Yourself

1. The quadratic equation $3(x+2)(x-p)=0$ is given. If one solution is $x=5$ what is the value of p?

2. If $(kx+3)(x-1)=0$, has solutions $x=-\frac{3}{2}$ and $x=1$.
What is the value of k?

3. One solution to $(x+5)(a-3x)=0$ is $x=\frac{1}{3}$.
What is the value of a?

4. Given the quadratic equation has solutions $x=-12, x=9$, and the equation is $ax^2+bx+c=0$. Find the value of a, b and c?

5. Given the quadratic equation has solutions $x=\frac{1}{4}, x=-3$, and the equation is $ax^2+bx+c=0$. Find the value of a, b and c?

6. Given the quadratic equation has solutions $x=\frac{2}{3}, x=\frac{1}{2}$, and the equation is $ax^2+bx+c=0$. Find the value of a, b and c?

Factorising: Null Factor Law and Factoring. Best Buds in Quadratics

If a quadratic can be factored.

With whole numbers.

Factor it first.

Then use the Null Factor Law to solve.

The factoring techniques are the same as before.

It'll be.

GCF.

Perfect squares.

DOTS.

or the AC Method.

If a quadratic doesn't factor nicely.

Solve them using the quadratic formula.

Which is covered next.

Here are the steps.

1) Write quadratic in standard form
2) Make one side equal zero
3) Factorise
4) Set each factor equal to zero and solve for the variable
5) Write both solutions with 'or' between them

By the way.

Everything we're working with here is a quadratic.

If you see any that look a bit odd.

Hint.

It has fractions with a variable in the denominator.

Multiply every term in the equation by that denominator.

That gets rid of it.

Simplify and the quadratic will look like a normal one again.

From there.

If it's in standard form.

One side equals zero.

You're good to go.

Factor.

Then solve.

If it's not in standard form.

Get it into standard form.

Get one side equal zero.

Factor.

Then solve.

Worked Example 1

Solve the following for x.

a $\mathbf{2x^2 - 4x = 0}$

b $\mathbf{15x - 12x^2 = 0}$

Explanation	**Solution**		
a Write the equation	$2x^2 - 4x = 0$		
Factor out the GCF.	$2x(x - 2) = 0$		
Solve each factor for x.	$2x = 0$	or	$x - 2 = 0$
	$x = 0$		$x = 2$
Write answer.	$x = 0 \ \ or \ \ x = 2$		
b Write the equation	$15x - 12x^2 = 0$		
Factor out the GCF.	$3x(5 - 4x) = 0$		
Solve each factor for x.	$3x = 0$	or	$5 - 4x = 0$
	$x = 0$		$4x = 5$
Write answer.	$x = 0 \ \ or \ \ x = \frac{5}{4}$		

Worked Example 2

Solve the following for x.

a $\mathbf{x^2 - 5x + 6 = 0}$

b $\mathbf{-2x^2 + 14x - 24 = 0}$

Explanation	Solution
a Write the equation.	$x^2 - 5x + 6 = 0$
Factorise using the AC Method.	$(x-2)(x-3) = 0$
Solve each factor for x.	$x - 2 = 0$ or $x - 3 = 0$ $x = 2$ $x = 3$
Write answer.	$x = 2$ *or* $x = 3$
b Write the equation.	$-2x^2 + 14x - 24 = 0$
Factor out the GCF.	$-2(x^2 - 7x + 12) = 0$
Factorise using the AC Method.	$-2(x-3)(x-4) = 0$
Divide both sides by number outside brackets.	$(x-3)(x-4) = 0$
Solve each factor for x.	$x - 3 = 0$ or $x - 4 = 0$ $x = 3$ $x = 4$
Write answer.	$x = 3$ *or* $x = 4$

Worked Example 3

Solve the following for x.

a $\mathbf{x^2 + 45 = 14x}$

b $\mathbf{5x - x^2 = -24}$

Explanation	**Solution**
a Write the equation.	$x^2 + 45 = 14x$
Rewrite in standard form.	$x^2 + 45 - 14x = 0$
Factorise using the AC Method.	$(x-5)(x-9) = 0$
Solve each factor for x.	$x - 5 = 0$ or $x - 9 = 0$ $x = 5$ $x = 9$
Write answer.	$x = 5 \ or \ x = 9$
b Write the equation.	$5x - x^2 = -24$
Rewrite in standard form.	$x^2 - 5x - 24 = 0$
Factorise using the AC Method.	$(x-8)(x+3) = 0$
Solve each factor for x.	$x - 8 = 0$ or $x + 3 = 0$ $x = 8$ $x = -3$
Write answer.	$x = 8 \ or \ x = -3$

Worked Example 4

Solve the following for x.

a $\quad \boldsymbol{x^2 - 2(5x - 8) = 0}$

b $\quad \boldsymbol{\dfrac{12 - x^2}{11} = x}$

Explanation	**Solution**
a Write the equation.	$x^2 - 2(5x - 8) = 0$
Expand the brackets.	$x^2 - 10x + 16 = 0$
Factorise using the AC Method.	$(x - 2)(x - 8) = 0$
Solve each factor for x.	$x - 2 = 0$ or $x - 8 = 0$ $x = 2$ $\quad$ $x = 8$
Write answer.	$x = 2 \quad or \quad x = 8$
b Write the equation.	$\dfrac{12 - x^2}{11} = x$
Move denominator to other side.	$12 - x^2 = 11x$
Rearrange into standard form.	$x^2 + 11x - 12 = 0$
Factorise using the AC Method.	$(x + 12)(x - 1) = 0$
Solve each factor for x.	$x + 12 = 0$ or $x - 1 = 0$ $x = -12$ $\quad$ $x = 1$
Write answer.	$x = -12 \quad or \quad x = 1$

Worked Example 5

Solve the following quadratic equations.

a $12a^2 = 6 - a$

b $2 - 7b = 15b^2$

Explanation	Solution
a Write the equation.	$12a^2 = 6 - a$
Rearrange into standard form.	$12a^2 + a + 6 = 0$
Factorise using the AC Method 2.	$(3a - 2)(4a + 3) = 0$
Solve each factor for x.	$3a - 2 = 0$ or $4a + 3 = 0$
	$a = \frac{2}{3}$ $a = -\frac{3}{4}$
Write answer.	$a = \frac{2}{3}$ or $a = -\frac{3}{4}$
b Write the equation.	$2 - 7b = 15b^2$
Rearrange into standard form.	$15b^2 + 7b - 2 = 0$
Factorise using the AC Method 2.	$(5b - 1)(3b + 2) = 0$
Solve each factor for x.	$5b - 1 = 0$ or $3b + 2 = 0$
	$b = \frac{1}{5}$ $b = -\frac{2}{3}$
Write answer.	$b = \frac{1}{5}$ or $b = -\frac{2}{3}$

Worked Example 6

Solve the quadratic equation: $(2w+1)(w-3)=-4w$

Explanation	Solution
Write the equation.	$(2w+1)(w-3)=-4w$
Expand brackets.	$2w^2-6w+w-3=-4w$
Combine like terms.	$2w^2-5w-3=-4w$
Move everything to one side	$2w^2-5w-3+4w=0$
Write in standard form.	$2w^2-w-3=0$
Factorise using the AC Method 2.	$(w+1)(2w-3)=0$
Solve each factor for x.	$w+1=0$ or $2w-3=0$ $w=-1$ $w=\frac{3}{2}$
Write answer.	$w=-1$ or $w=\frac{3}{2}$

Worked Example 7

Solve the quadratic equation: $\mathbf{2 - 3m = \frac{1}{3m}}$

Explanation	Solution
Write the equation.	$2 - 3m = \dfrac{1}{3m}$
Move denominator to the left.	$3m(2 - 3m) = 1$
Expand the brackets.	$6m - 9m^2 = 1$
Move everything to one side	$9m^2 - 6m + 1 = 0$
Factorise the Perfect Square.	$9m^2 - 6m + 1 = 0$
Take square root of both sides.	$(3m - 1)^2 = 0$
Solve.	$3m - 1 = 0$
	$3m = 1$
	$m = \dfrac{1}{3}$

Not making the equation equal zero first before solving. ⚠

Worked Example 8

Solve the quadratic equation: $\dfrac{15}{x-1} = x - 3$

Explanation	Solution
Write the equation.	$\dfrac{15}{x-1} = x - 3$
Move denominator to the right.	$15 = (x-3)(x-1)$
Expand brackets.	$15 = x^2 - 4x + 3$
Move everything to one side	$x^2 - 4x - 12 = 0$
Factorise using the AC Method.	$(x-6)(x+2) = 0$
Solve each factor for x.	$x - 6 = 0$ or $x + 2 = 0$ $x = 6$ $\quad$ $x = -2$
Write answer.	$x = 6 \quad or \quad x = -2$

Not multiplying every term in the equation by the denominator to get rid of the fraction.

Worked Example 9

Solve the equation: $\dfrac{b}{b-3}-\dfrac{b+2}{b+5}=1$

Explanation	Solution
Write the equation.	$\dfrac{b}{b-3}-\dfrac{b+2}{b+5}=1$
Combine fractions on the left.	$\dfrac{b(b+5)-(b+2)(b-3)}{(b-3)(b+5)}=1$
Move denominator to the right.	$b(b+5)-(b+2)(b-3)=(b-3)(b+5)$
Expand brackets on each side.	$b^2+5b-\left(b^2-b-6\right)=b^2+2b-15$
	$b^2+5b-b^2+b+6=b^2+2b-15$
Combine like terms.	$6b+6=b^2+2b-15$
Write in standard form.	$b^2-4b-21=0$
Factorise using the AC Method.	$(b-7)(b+3)=0$
Solve each factor for x.	$b-7=0$ or $b+3=0$ $b=7$ $b=-3$
Write answer.	$b=7 \;\; or \;\; b=-3$

Not being careful when expanding brackets with a minus in front.
The minus changes all the signs inside the brackets.

Worked Example 10

Find the value of k for the following.

a **The quadratic $3x^2 + 7x + k = 0$ has $x = -3$ as a solution.**

b **The quadratic $2x^2 - kx + 3 = 0$ has $(x - 1)$ as one factor.**

	Explanation	**Solution**
a	Write the equation.	$3x^2 + 7x + k = 0$
	Use the given solution.	$x = -3$
	Substitute the solution in the quadratic.	$3(-3)^2 + 7(-3) + k = 0$
	Simplify.	$27 - 21 + k = 0$
	Solve for k.	$6 + k = 0$
		$k = -6$
b	Write the equation.	$2x^2 - kx + 3 = 0$
	Let the factor equal 0 and solve for x.	$x \quad 1 = 0$ $x = 1$
	Substitute the solution in the quadratic.	$2(1)^2 - k(1) + 3 = 0$
	Simplify.	$2 - k + 3 = 0$
	Solve for k.	$k = 5$

When you substitute a solution back into a quadratic it will equal zero.
Use this to solve for an unknown constant.

Worked Example 11

The quadratic $5x^2 - kx + 2 = 0$ has $(5x - 2)$ as one factor. Find k?

Explanation	**Solution**
Write the second equation.	$5x^2 - kx + 2 = 0$
Let the factor equal 0 and solve for x.	$5x - 2 = 0$
	$x = \frac{2}{5}$
Substitute the solution in the quadratic.	$5\left(\frac{2}{5}\right)^2 - k\left(\frac{2}{5}\right) + 2 = 0$
Simplify.	$5 \times \frac{4}{25} - \frac{2k}{5} + 2 = 0$
	$\frac{4}{5} - \frac{2k}{5} + 2 = 0$
Combine fractions.	$\frac{4 - 2k}{5} + 2 = 0$
Move number to other side.	$\frac{4 - 2k}{5} = -2$
Move denominator to other side.	$4 - 2k = -10$
Solve for k.	$14 = 2k$
	$k = 7$

Your Turn To Try

1 **What are the solutions to the equation $4x^2 + 12x = 0$?**

A $x = -3\ only$

B $x = 0\ or\ x = 3$

C $x = 0\ or\ x = -6$

D $x = 0\ or\ x = -3$

2 **Solve the equation $x^2 + 10x + 25 = 0$. Which of the following is correct?**

A $x = -5$

B $x = 0\ or\ x = -5$

C $x = 5$

D $x = 0\ or\ x = 5$

3 **Which of the following are solutions to the equation $x^2 - 10x + 16 = 0$?**

A $x = 4\ only$

B $x = 2\ or\ x = 8$

C $x = 4\ or\ x = -4$

D $x = -2\ or\ x = -8$

4 **Which equation has solutions $x = -4\ or\ x = 1$?**

A $x^2 + 4x - 3 = 0$

B $x^2 - 3x - 4 = 0$

C $x^2 - 5x - 4 = 0$

D $x^2 + 3x - 4 = 0$

5 **If $x^2 - 45 = 4x$, what are the solutions?**

A $x = 9\ or\ x = -5$

B $x = 0\ or\ x = 4$

C $x = -9\ or\ x = 5$

D $x = 0\ or\ x = -4$

Question	1	2	3	4	5
Answer	D	A	B	D	A

Time To Practice

1 Solve the following quadratic equations by factorising.

a $x^2 + 7x = 0$
b $a^2 - 5a = 0$
c $b^2 - 10b = 0$

d $3x^2 - 9x = 0$
e $5a^2 - 10a = 0$
f $-b^2 - b = 0$

g $36x = 6x^2$
h $2a^2 = 22a$
i $-8b^2 = 24b$

2 Solve the following quadratic equations.

a $x^2 + 5x + 6 = 0$
b $a^2 - 7a + 10 = 0$
c $2b^2 + 8b + 6 = 0$

d $2x^2 - 4x = 6$
e $3a^2 = 11a - 10$
f $7b - 2b^2 = 3$

g $10x - 3x^2 = 8$
h $4a^2 = 3 + 4a$
i $15 = 11b - 2b^2$

3 Solve the following quadratic equations.

a $4x^2 - 5(4x - 5) = 0$
b $4p^2 + 3(4p + 3) = 0$

c $x(x + 4) = 4x + 9$
d $4(u^2 + 2) - 9 = 0$

e $s^2 + s - 9 = 5s - 4$
f $3y^2 - 4(2y - 1) = 0$

g $3(1 - 2a^2) - 17a = 0$
h $9x^2 - (x - 1)^2 = 0$

i $3t(t + 6) = 4(t - 2)$
j $6k(k + 1) = 15 - 4k + k^2$

4 Solve the following by first writing in the form $ax^2 + bx + c = 0$.

a $\dfrac{5x + 84}{x} = x$
b $\dfrac{3}{v} = v + 2$

c $\dfrac{4}{x - 2} = x + 1$
d $\dfrac{1}{h} = 3 - 2h$

e $\dfrac{20 - 3a}{a} = 2a$
f $\dfrac{7w + 10}{2w} = 3w$

Challenge Yourself

1 Solve these equations by first multiplying by an appropriate expression.

a $12x = 13 - \frac{3}{x}$

b $\frac{10}{x} - 21 = 10x$

c $\frac{x+4}{2} - \frac{3}{x-3} = 1$

d $\frac{x}{x-2} - \frac{x+1}{x+4} = 1$

2 Given the quadratic equation has solutions $x = -7, x = p$, and the equation is $x^2 - x - 56 = 0$. Find the value of p?

3 The quadratic equation $x^2 - px + 10 = 0$ in factored form is $(x - a)(x - b) = 0$. If $a = 2$, what is the value of b and p?

4 The quadratic equation $x^2 + 10x + k = 0$ has **exactly one** solution. What is the value of k and what is the solution?

5 For what value of a does the quadratic equation $ax^2 + 9x + 2 = 0$ have $x = -2$ as a solution.

6 For what value of c does the quadratic equation $3x^2 + 7x + c = 0$ have $x = -3$ as a solution.

The Quadratic Formula: Meet Algebra's Most Famous Formula

The quadratic formula can be used to solve any quadratic.

But it's best used for ones that don't factor nicely.

Just plug in the numbers and solve.

This is the formula.

$$x = \frac{-b \pm \sqrt{b^2 - 4ac}}{2a}$$

The values a, b and c come from the standard form.

- a is the number in front of the squared term
- b is the number in front of the middle term
- c is the number by itself

Replace these values into the formula and simplify.

No need for any factoring.

Here are the steps.

1) Get the quadratic in standard form (equal to zero if it isn't already)
2) Write down the values of a, b and c
3) Replace the letters with the numbers in the formula
4) Simplify
5) If the square root becomes a whole number, split it into two answers
    I. If it doesn't, just leave the answer as one expression
6) Simplify each answer
7) Write both solutions with 'or' between them

Worked Example 1

Use the quadratic formula to solve $x^2 - 10x + 16 = 0$.

Explanation	**Solution**
Write the equation.	$x^2 - 10x + 16 = 0$
Write the values of a, b and c.	$a = 1, b = -10, c = 16$
	$x = \dfrac{-b \pm \sqrt{b^2 - 4ac}}{2a}$
Substitute these values into the formula.	$x = \dfrac{-(-10) \pm \sqrt{(-10)^2 - 4(1)(16)}}{2(1)}$
Simplify.	$= \dfrac{10 \pm \sqrt{100 - 64}}{2}$
	$= \dfrac{10 \pm \sqrt{36}}{2}$
	$= \dfrac{10 \pm 6}{2} = \dfrac{5 \pm 3}{1}$
	$= 5 \pm 3$
Split it into the two answers.	$x = 5 - 3 = 2$ *or* $x = 5 + 3 = 8$
Simplify each answer.	$x = 2$, $x = 8$

When substituting values, put them in brackets.
Helps avoid mistakes with the negatives.

Worked Example 2

Use the quadratic formula to solve $x^2 - 5x + 6 = 0$.

Explanation	Solution
Write the equation.	$x^2 + 5x + 6 = 0$
Write the values of a, b and c.	$a = 1, b = -5, c = 6$
	$x = \dfrac{-b \pm \sqrt{b^2 - 4ac}}{2a}$
Substitute these values into the formula.	$x = \dfrac{-(-5) \pm \sqrt{(-5)^2 - 4(1)(6)}}{2(1)}$
Simplify.	$= \dfrac{5 \pm \sqrt{25 - 24}}{2}$
	$= \dfrac{5 \pm \sqrt{1}}{2}$
	$= \dfrac{5 \pm 1}{2}$
Split it into the two answers.	$x = \dfrac{5 - 1}{2}$ *or* $x = \dfrac{5 + 1}{2}$
Simplify each answer.	$= \dfrac{4}{2}$ $\quad = \dfrac{6}{2}$
	$x = 2$ $\quad x = 3$

Forgetting to include the minus in front of b when b is negative.

Worked Example 3

Use the quadratic formula to solve $2x^2 + 5x - 12 = 0$.

Explanation	Solution
Write the equation.	$2x^2 + 5x - 12 = 0$
Write the values of a, b and c.	$a = 2, b = 5, c = -12$
	$x = \dfrac{-b \pm \sqrt{b^2 - 4ac}}{2a}$
Substitute these values into the formula.	$x = \dfrac{-5 \pm \sqrt{(5)^2 - 4(2)(-12)}}{2(2)}$
Simplify.	$= \dfrac{-5 \pm \sqrt{25 + 96}}{4}$
	$= \dfrac{-5 \pm \sqrt{121}}{4}$
	$= \dfrac{-5 \pm 11}{4}$
Split it into the two answers.	$x = \dfrac{-5 - 11}{4}$ *or* $x = \dfrac{-5 + 11}{4}$
Simplify each answer.	$= \dfrac{-16}{4}$ $\quad$ $= \dfrac{6}{4}$
	$x = -4$ $\quad$ $x = \dfrac{3}{2}$

Worked Example 4

Use the quadratic formula to solve $-2x^2 + 2x + 1 = 0$.

Explanation	**Solution**
Write the equation.	$-2x^2 + 2x + 1 = 0$
Write the values of a, b and c.	$a = -2, b = 2, c = 1$
	$x = \dfrac{-b \pm \sqrt{b^2 - 4ac}}{2a}$
Substitute these values into the formula.	$x = \dfrac{-(2) \pm \sqrt{(2)^2 - 4(-2)(1)}}{2(-2)}$
Simplify.	$= \dfrac{-2 \pm \sqrt{4+8}}{-4}$
	$= \dfrac{2 \pm \sqrt{12}}{4}$
	$= \dfrac{2 \pm \sqrt{4 \times 3}}{4}$
	$= \dfrac{2 \pm 2\sqrt{3}}{4}$
Write final answer.	$x = \dfrac{1 \pm \sqrt{3}}{2}$

Always simplify the answer if you can. Simplify the square root and cancel out any common factors.

Worked Example 5

Use the quadratic formula to solve $10x + 2 = -5x^2$.

Explanation	**Solution**
Write the equation.	$10x + 2 = -5x^2$
Rearrange and write in standard form.	$5x^2 + 10x + 2 = 0$
Write the values of a, b and c.	$a = 5, b = 10, c = 2$
	$x = \frac{-b \pm \sqrt{b^2 - 4ac}}{2a}$
Substitute these values into the formula.	$x = \frac{-(10) \pm \sqrt{(10)^2 - 4(5)(2)}}{2(5)}$
Simplify.	$= \frac{-10 \pm \sqrt{100 - 40}}{10}$ $= \frac{-10 \pm \sqrt{60}}{10}$ $= \frac{-10 \pm 2\sqrt{15}}{10}$
Simplify final answer.	$x = \frac{-5 \pm \sqrt{15}}{5}$

Canceling numbers that don't go into ALL three parts.
The denominator, the number before the $\pm$ signs, and the number in front of the square root.

Worked Example 6

Use the quadratic formula to solve $(2x+1)(x+5)=1$.

Explanation	Solution
Write the equation.	$(2x+1)(x+5)=1$
Expand the brackets.	$2x^2+11x+5=1$
Rearrange and write in standard form.	$2x^2+11x+4=0$
Write the values of a, b and c.	$a=2, b=11, c=4$
	$x=\dfrac{-b\pm\sqrt{b^2-4ac}}{2a}$
Substitute these values into the formula.	$x=\dfrac{-(11)\pm\sqrt{(11)^2-4(2)(4)}}{2(2)}$
Simplify.	$=\dfrac{-11\pm\sqrt{121-32}}{4}$
Write final answer.	$x=\dfrac{-11\pm\sqrt{89}}{4}$

Writing 2 at the bottom instead of $2a$. (2 multiplied by a). ⚠

Worked Example 7

Use the quadratic formula to solve $\frac{1}{3}x^2 - \frac{8}{3}x + 2 = 0$.

Explanation	Solution
Write the quadratic.	$\frac{1}{3}x^2 - \frac{8}{3}x + 2 = 0$
Multiply both sides by 3 to get of fractions.	$x^2 - 8x + 6 = 0$
Identify the values of a, b and c.	$a = 1, b = -8, c = 6$
	$x = \frac{-b \pm \sqrt{b^2 - 4ac}}{2a}$
Substitute these values into the formula.	$x = \frac{-(-8) \pm \sqrt{(-8)^2 - 4(1)(6)}}{2(1)}$
Simplify.	$= \frac{8 \pm \sqrt{64 - 24}}{2} = \frac{8 \pm \sqrt{40}}{2}$
	$= \frac{8 \pm \sqrt{4 \times 10}}{2}$
	$= \frac{8 \pm 2\sqrt{10}}{2} = \frac{4 \pm \sqrt{10}}{1}$
Write final answer.	$x = 4 \pm \sqrt{10}$

Get rid of fractions by multiplying every term - on both sides - by the denominator. One side being zero will remain zero.

Worked Example 8

Solve the following equations for x in terms of k.

a $\mathbf{2x^2 - kx + 4 = 0}$

b $\mathbf{2x^2 - 3x - k = 0}$

Explanation	Solution
a Write the equation.	$2x^2 - kx + 4 = 0$
Write the values of a, b and c.	$a = 2, b = -k, c = 4$
	$x = \dfrac{-b \pm \sqrt{b^2 - 4ac}}{2a}$
Substitute these values into the formula.	$x = \dfrac{-(-k) \pm \sqrt{(-k)^2 - 4(2)(4)}}{2(2)}$
Simplify.	$x = \dfrac{k \pm \sqrt{k^2 - 32}}{4}$
b Write the equation.	$2x^2 - 3x - k = 0$
Write the values of a, b and c.	$a = 2, b = -3, c = -k$
	$x = \dfrac{-b \pm \sqrt{b^2 - 4ac}}{2a}$
Substitute these values into the formula.	$x = \dfrac{-(-3) \pm \sqrt{(-3)^2 - 4(2)(-k)}}{2(2)}$
Simplify.	$x = \dfrac{3 \pm \sqrt{9 + 8k}}{4}$

Worked Example 9

Use the quadratic formula to solve $28x - 4 - 49x^2 = 0$.

Explanation	Solution
Write the equation in standard form.	$-49x^2 + 28x - 4 = 0$
Write the values of a, b and c.	$a = -49, b = 28, c = -4$
	$x = \dfrac{-b \pm \sqrt{b^2 - 4ac}}{2a}$
Substitute these values into the formula.	$x = \dfrac{-28 \pm \sqrt{(28)^2 - 4(-49)(-4)}}{2(-49)}$
Simplify.	$= \dfrac{-28 \pm \sqrt{784 - 784}}{-98}$
	$= \dfrac{-28 \pm \sqrt{0}}{-98} = \dfrac{-28}{-98}$
Write final answer.	$= \dfrac{28}{98} = \dfrac{14 \times 2}{14 \times 7}$
	$x = \dfrac{2}{7}$

Worked Example 10

Use the quadratic formula to solve $4x^2 + x + 2 = 0$.

Explanation	Solution
Write the equation.	$4x^2 + x + 2 = 0$
Identify the values of a, b and c.	$a = 4, b = 1, c = 2$
	$x = \dfrac{-b \pm \sqrt{b^2 - 4ac}}{2a}$
Substitute these values into the formula.	$x = \dfrac{-1 \pm \sqrt{(1)^2 - 4(4)(2)}}{2(8)}$
Simplify.	$= \dfrac{-1 \pm \sqrt{1 - 31}}{16}$
	$x = \dfrac{-1 \pm \sqrt{-31}}{16}$
	There are no solutions because we can't take the square root of a negative number.
Write final answer.	No Solution.

Taking the square root of a negative number as if it were positive.

The Discriminant: Meet Delta. The Little Formula Inside The Big Formula

There's a quick way to tell how many solutions a quadratic has.

Without even solving it.

By using the *Discriminant*.

$$\Delta = b^2 - 4ac$$

It's the part under the square root in the quadratic formula.

Not the whole square root.

Just what's underneath.

And it's represented by the symbol Δ.

The Greek letter Delta.

Here's how it works.

Substitute the a, b, and c values and work it out.

If,

$\Delta < 0$ → No solution

$\Delta = 0$ → One solution

$\Delta > 0$ → Two solutions

The discriminant is useful when graphing quadratics.

The number of solutions is the number of times the graph will cross the x-axis.

Later on.

It's used to check if a straight-line graph intersects a quadratic graph.

For now, just the basics.

Finding out how many solutions a quadratic equation has.

Here are the steps.

1) Write equation in standard form
2) Write down a, b and c
3) Substitute these values into the discriminant formula
4) Work it out
5) Comment on the number of solutions

Worked Example 11

Find the value of the discriminant for the following equations.

a $\mathbf{3x^2 - 5x + 2 = 0}$ b $\mathbf{-2x^2 + 12x - 7 = 0}$

Explanation		**Solution**
a	Write the quadratic in standard form.	$3x^2 - 5x + 2 = 0$
	Identify the values of a, b and c.	$a = 3, b = -5, c = 2$
	Write the formula for Δ.	$\Delta = \boldsymbol{b^2 - 4ac}$
	Substitute the values $a, b \ and \ c$.	$\Delta = (-5)^2 - 4(3)(2)$
	Simplify.	$= 25 - 24$
		$= 1$
b	Write the quadratic in standard form.	$-2x^2 + 12x - 7 = 0$
	Identify the values of a, b and c.	$a = -2, b = 12, c = -7$
	Write the formula for Δ.	$\Delta = \boldsymbol{b^2 - 4ac}$
	Substitute the values $a, b \ and \ c$.	$\Delta = (12)^2 - 4(-2)(-7)$
	Simplify.	$= 144 - 56$
		$= 88$

Including the square root sign when working out the discriminant.

Worked Example 12

How many solutions are there to the following equations.

a $\mathbf{5x^2 + 9x - 2 = 0}$

b $\mathbf{12x^2 - 3x + 1 = 0}$

Explanation	Solution
a Write the quadratic in standard form.	$5x^2 + 9x - 2 = 0$
Identify the values of a, b and c.	$a = 5, b = 9, c = -2$
Write the formula for Δ.	$\Delta = \boldsymbol{b^2 - 4ac}$
Substitute the values a, b and c.	$\Delta = (9)^2 - 4(5)(-2)$
Simplify.	$= 81 + 40$
	$= 121$
Comment on the number of solutions.	Since $\Delta > 0$, there are two solutions.
b Write the quadratic in standard form.	$12x^2 - 3x + 1 = 0$
Identify the values of a, b and c.	$a = 12, b = -3, c = 1$
Write the formula for Δ.	$\Delta = \boldsymbol{b^2 - 4ac}$
Substitute the values a, b and c.	$\Delta = (-3)^2 - 4(12)(1)$
Simplify.	$= 9 - 48$
	$= -39$
Comment on the number of solutions.	Since $\Delta < 0$, there are no solutions.

Worked Example 13

How many solutions are there to the following equations.

a $\mathbf{4x^2 - 20x + 25 = 0}$ b $\mathbf{-7x^2 + 6x - 3 = 0}$

Explanation	**Solution**
a Write the quadratic in standard form.	$4x^2 - 20x + 25 = 0$
Identify the values of a, b and c.	$a = 4, b = -20, c = 25$
Write the formula for Δ.	$\Delta = \boldsymbol{b^2 - 4ac}$
Substitute the values a, b and c.	$\Delta = (-20)^2 - 4(4)(25)$
Simplify.	$= 400 - 400$
	$= 0$
Comment on the number of solutions.	Since $\Delta = 0$, there is one solution.
b Write the quadratic in standard form.	$-7x^2 + 6x - 3 = 0$
Identify the values of a, b and c.	$a = -7, b = 6, c = -3$
Write the formula for Δ.	$\Delta = \boldsymbol{b^2 - 4ac}$
Substitute the values a, b and c.	$\Delta = (6)^2 - 4(-7)(-3)$
Simplify.	$= 36 - 84$
	$= -48$
Comment on the number of solutions.	Since $\Delta < 0$, there are no solutions.

Worked Example 14

For what values of k does $x^2 + kx + 9 = 0$ have only one solution?

Explanation	Solution
Write the equation.	$x^2 + kx + 9 = 0$
Identify the values of a, b and c.	$a = 1, b = k, c = 9$
Write the formula for Δ.	$\Delta = \boldsymbol{b^2 - 4ac}$
Substitute the values a, b and c.	$\Delta = (k)^2 - 4(1)(9)$
Simplify.	$= k^2 - 36$
Let $\Delta = 0$ for one solution.	Let $\Delta = 0$
Solve for k.	$k^2 - 36 = 0$
	$k^2 = 36$
	$k = \pm\sqrt{36}$
	$k = \pm 6$

Your Turn To Try

1 **If the discriminant is negative how many real solution(s) does the quadratic equation have?**

A 1　　B 2

C 0　　D $All\ of\ the\ above$

2 **Which of the following values of a, b and c will result in a positive discriminant?**

A $a = 2, b = 4, c = 6$　　B $a = 1, b = -3, c = 2$

C $a = 1, b = 2, c = 5$　　D $a = 5, b = 4, c = 1$

3 **Given the equation $3x^2 - 6x + 2 = 0$, what is the value of the discriminant?**

A 12　　B 13

C $\sqrt{12}$　　D $\sqrt{13}$

4 **In the quadratic equation $-2x^2 = 7x - 1$, what are the values of a, b and c?**

A $a = -2, b = 7, c = -1$　　B $a = 2, b = -7, c = -1$

C $a = 2, b = 7, c = 1$　　D $a = -2, b = -7, c = 1$

5 **Which of the following equations CANNOT be solved using the quadratic formula?**

A $x^2 - 4 = 0$　　B $x + 5 = 0$

C $3x^2 - 7x + 4 = 0$　　D $x^2 = 8x - 6$

Question	1	2	3	4	5
Answer	C	B	A	D	B

Time To Practice

1 For the following quadratic equations find the value of the discriminant.

a $x^2 + 5x + 3 = 0$ b $x^2 + 6x + 9 = 0$ c $2x^2 - 3x + 4 = 0$

d $3x^2 + 2x + 1 = 0$ e $5x^2 + 3x - 2 = 0$ f $-3x^2 + 4x + 5 = 0$

2 Determine the number of solutions to the following quadratic equations using the discriminant.

a $x^2 + 7x - 1 = 0$ b $x^2 + 4x + 5 = 0$ c $2x^2 - 6x + 3 = 0$

d $-2x^2 + 3x - 1 = 0$ e $4x^2 - 5x + 2 = 0$ f $9x^2 + 6x + 1 = 0$

3 Find the solutions to the following using the quadratic formula.

a $x^2 + 5x - 6 = 0$ b $x^2 - 6x + 8 = 0$ c $x^2 - 8x + 16 = 0$

d $-x^2 + 5x + 14 = 0$ e $-x^2 + 8x + 9 = 0$ f $x^2 - 12x + 36 = 0$

g $-x^2 + 11x - 30 = 0$ h $25x^2 + 20x + 4 = 0$ i $4x^2 - 4x + 1 = 0$

4 Solve the following using the quadratic formula.

a $2x^2 + x - 3 = 0$ b $2x^2 - 9x + 4 = 0$ c $2x^2 + 5x + 3 = 0$

d $4x^2 - 8x + 3 = 0$ e $12x^2 - 7x + 1 = 0$ f $9x^2 - 9x + 2 = 0$

g $3x^2 - x - 10 = 0$ h $12x^2 + 4x - 1 = 0$ i $5x^2 - 11x + 2 = 0$

5 Solve for x in each of the following using the quadratic formula.

a $x^2 + x - 1 = 0$ b $x^2 + 5x + 3 = 0$ c $x^2 + 3x - 2 = 0$

d $-x^2 - 3x + 6 = 0$ e $-x^2 + 3x + 3 = 0$ f $2x^2 + x - 5 = 0$

g $-2x^2 - 5x + 4 = 0$ h $3x^2 - 9x + 5 = 0$ i $3x^2 - 11x + 9 = 0$

Time To Practice

6 Solve the following using the quadratic formula.

a $2x^2 = 3 - 4x$

b $3x^2 = 1 + 6x$

c $10x = 4x^2 + 3$

d $-3 - 2x = -2x^2$

e $-7 = -4x^2 - 4x$

f $3x^2 = 5 + 6x$

g $6x = 8x^2 - 1$

h $-2x - 3 = -6x^2$

i $5x^2 - 4 = -2x$

7 Solve the following using the quadratic formula.

a $(2x + 1)(x - 3) = 3$

b $(2 - x)(3x + 1) = -1$

c $5 - 2(x^2 + 2) = 2x$

d $(3x - 2)(x + 1) = 4x + 7$

e $(x + 2)(2x - 1) = 5 - x$

f $(5 - 2x)(x - 1) = -(x + 3)$

8 Solve the following using the quadratic formula.

a $\frac{1}{2}x^2 + 3x - 5 = 0$

b $\frac{3}{4}x^2 + 2x - 1 = 0$

c $x^2 - \frac{1}{3}x - 1 = 0$

d $3x^2 - \frac{1}{2}x - 2 = 0$

e $-x^2 - \frac{2}{3}x + 5 = 0$

f $\frac{1}{2}x^2 + \frac{2}{3}x - \frac{5}{6} = 0$

g $5x = \frac{2x + 2}{x}$

h $\frac{3}{x} = 3x + 4$

i $3x = \frac{10x - 1}{2x}$

Challenge Yourself

1 Solve the following equations for x in terms of k, where k is a constant.

a $x^2 - kx + 8 = 0$

b $kx^2 + 5x - 7 = 0$

c $3x^2 + 4x - k = 0$

d $-2x^2 - 3kx + 2k^2 = 0$

2 If the discriminant of $x^2 - 4x + c = 0$ is -12. What is the value of c?

3 Find the value of b such that the discriminant of $5 + bx - 2x^2 = 0$ equals 56.

4 Find the value of k that makes the following quadratic equation have exactly one real solution: $3x^2 + kx + 12 = 0$.

5 For what values of k, where k is a constant will the equation $x^2 + (k - 3)x + 3 - 2k = 0$ have exactly one solution?

3. Answers

Square Root Method

Time to Practice

1	a	$x = \pm 4$	b	$a = \pm 8$	c	$b = \pm 5$
	d	$x = \pm 11$	e	$a = \pm 3$	f	$b = \pm 1$
	g	$x = \pm 10$	h	$a = \pm 6$	i	$b = \pm\sqrt{7}$
	j	*No Solution*	k	*No Solution*	l	*No Solution*
	m	$x = \pm\sqrt{10}$	n	$a = \pm\sqrt{15}$	o	$b = \pm\sqrt{11}$
2	a	$x = \pm\frac{1}{2}$	b	$a = \pm\frac{3}{8}$	c	$b = \pm\frac{1}{6}$
	d	$x = \pm\frac{2}{3}$	e	$a = \pm\frac{7}{6}$	f	$b = \pm\frac{10}{9}$
3	a	$x = -4$	b	$a = 5,$	c	$b = 2,$
	d	$x = -\frac{3}{2}$	e	$a = \frac{7}{6}$	f	$b = -\frac{1}{4}$
	g	$x = -6, x = 0$	h	$a = 1, a = 9$	i	$b = 13, b = 1$
4	a	$x = -3, x = 1$	b	$a = -5, a = -1$	c	$b = 2, b = 8$
	d	$x = 2, x = 6$	e	$a = 4, a = 0$	f	$b = \frac{1}{3}, b = \frac{7}{3}$
	g	$x = -3, x = 2$	h	$a = \frac{9}{2}, a = \frac{5}{2}$	i	$b = 0, b = \frac{5}{2}$
	j	$x = 0, x = \frac{10}{3}$	k	$a = \frac{5}{2}, a = -\frac{3}{2}$	l	$b = -\frac{2}{5}, b = \frac{4}{5}$

Challenge Yourself

1	a	$x = \frac{1 \pm 2\sqrt{2}}{2}$	b	$m = \frac{1 \pm 3\sqrt{2}}{2}$	c	$x = \frac{1 \pm \sqrt{15}}{2}$
	d	$n = \frac{1 \pm 2\sqrt{3}}{2}$	e	$p = \frac{-2 \pm 2\sqrt{6}}{3}$	f	$x = \frac{-4 \pm 2\sqrt{7}}{3}$
	g	$x = \frac{1 \pm 9\sqrt{2}}{3}$	h	$q = \frac{-1 \pm 6\sqrt{5}}{2}$	2	$k = \pm 3$

Completing The Square

Time to Practice

1 a $x = -1 \pm \sqrt{3}$ b $a = 1 \pm \sqrt{6}$ c $b = -1 \pm \sqrt{2}$

d $x = 1 \pm \sqrt{5}$ e $a = -3 \pm \sqrt{2}$ f $b = 4 \pm \sqrt{7}$

g $x = -3 \pm \sqrt{19}$ h $a = 5 \pm \sqrt{29}$ i $b = -6 \pm \sqrt{35}$

2 a $x = \frac{1 \pm \sqrt{5}}{2}$ b $a = \frac{-3 \pm \sqrt{17}}{2}$ c $b = \frac{5 \pm \sqrt{37}}{2}$

d $x = \frac{3 \pm \sqrt{5}}{2}$ e $a = \frac{-5 \pm \sqrt{13}}{2}$ f $b = \frac{-7 \pm \sqrt{53}}{2}$

g $x = \frac{-3 \pm \sqrt{33}}{2}$ h $a = \frac{9 \pm \sqrt{53}}{2}$ i $b = \frac{11 \pm \sqrt{41}}{2}$

j $x = \frac{5 \pm \sqrt{65}}{2}$ k $a = \frac{11 \pm \sqrt{21}}{2}$ l $b = \frac{-7 \pm \sqrt{13}}{2}$

3 a $x = -1 \pm \sqrt{5}$ b $a = -2 \pm \sqrt{6}$ c $b = 1 \pm \sqrt{3}$

d $x = \frac{3 \pm \sqrt{21}}{2}$ e $a = \frac{3 \pm \sqrt{5}}{2}$ f $b = \frac{3 \pm \sqrt{13}}{2}$

g $x = \frac{-3 \pm \sqrt{29}}{2}$ h $a = \frac{-1 \pm \sqrt{13}}{2}$ i $b = \frac{-9 \pm \sqrt{5}}{2}$

4 a $x = \frac{1 \pm \sqrt{21}}{5}$ b $a = \frac{-1 \pm \sqrt{13}}{3}$ c $b = \frac{1 \pm \sqrt{21}}{4}$

d $x = \frac{5 \pm \sqrt{61}}{6}$ e $a = \frac{2 \pm \sqrt{10}}{3}$ f $b = \frac{1 \pm \sqrt{6}}{5}$

g $x = \frac{7 \pm \sqrt{73}}{6}$ h *No Solution* i *No Solution*

Challenge Yourself

1 a $x = \frac{-4 \pm 3\sqrt{5}}{2}$ b $a = \frac{-9 \pm 2\sqrt{7}}{3}$ c $x = \frac{10 \pm 3\sqrt{11}}{4}$

d $x = \frac{-6 \pm \sqrt{6}}{9}$

2 $p = 3, q = -11$

3 $b = 12$

4 $m = -3$

5 $a = -3, b = 1, c = 7$

6 $a = 2, b = -3, c = 3$

Null Factor Law

Time to Practice

1 a $x = 0, x = -3$ b $a = 0, a = -5$ c $b = 0, b = 7$
d $x = 0, x = -2$ e $a = 0, a = 3$ f $b = 0, b = 9$
g $x = 0, x = 4$ h $a = 0, a = 1$ i $b = 0, b = -8$

2 a $x = 5, x = -3$ b $a = 4, a = -1$ c $b = 2, b = 6$
d $x = 4, x = 8$ e $a = 8, a = 3$ f $b = 3, b = 9$
g $x = 4, x = -3$ h $a = 2, a = 9$ i $b = 5, b = 6$

3 a $x = -\frac{1}{3}, x = 4$ b $m = \frac{5}{2}, m = -6$ c $p = 2, p = \frac{3}{5}$
d $a = -\frac{1}{4}, a = \frac{7}{2}$ e $u = -\frac{5}{2}, u = 1$ f $w = 3, w = -\frac{4}{3}$
g $x = \frac{5}{3}, x = -\frac{2}{3}$ h $y = \frac{7}{2}, y = \frac{4}{3}$ i $w = -\frac{3}{4}, w = \frac{9}{2}$
j $t = \frac{1}{7}, t = \frac{11}{2}$

Challenge Yourself

1 $p = 5$ 2 $k = 2$ 3 $a = 1$
4 $a = 1, b = 3, c = -108$ 5 $a = 4, b = 11, c = -3$ 6 $a = 6, b = -7, c = 2$

Factorising

Time to Practice

1 a $x = 0, x = -7$ b $a = 0, a = 5$ c $b = 0, b = 10$

d $x = 0, x = 3$ e $a = 0, a = 2$ f $b = 0, b = -1$

g $x = 0, x = 6$ h $a = 0, a = 11$ i $b = 0, b = -3$

2 a $x = -2, x = -3$ b $a = 5, a = 2$ c $b = -1, b = -3$

d $x = 3, x = -1$ e $a = \frac{5}{3}, a = 2$ f $b = \frac{1}{2}, b = 3$

g $x = \frac{4}{3}, x = 2$ h $a = -\frac{1}{2}, a = \frac{3}{2}$ i $b = \frac{5}{2}, b = 3$

3 a $x = \frac{5}{2}$ b $p = -\frac{3}{2}$ c $x = \pm 3$

d $u = \pm\frac{1}{2}$ e $s = 5, s = -1$ f $y = \frac{2}{3}, y = 2$

g $a = \frac{1}{6}, a = -3$ h $x = \frac{1}{4}, x = \frac{1}{2}$ i $t = -\frac{2}{3}, t = -4$

j $k = -3, k = 1$

4 a $x = 12, x = -7$ b $v = 1, v = -3$ c $x = 3, x = -2$

d $h = \frac{1}{2}, h = 1$ e $a = \frac{5}{2}, a = -4$ f $w = -\frac{5}{6}, w = 2$

Challenge Yourself

1 a $x = \frac{1}{3}, x = \frac{3}{4}$ b $x = \frac{2}{5}, x = -\frac{5}{2}$ c $x = 4, x = -3$

d $x = 5, x = -2$

2 $p = 8$ 3 $b = 5, p = 7$ 4 $k = 25 \text{ and } x = -5$

5 $a = 4$ 6 $c = -6$

The Quadratic Formula

Time to Practice

1 a 13 b 0 c −23
d −8 e 49 f 76

2 a $2\,Solutions$ b $0\,Solutions$ c $2\,Solutions$
d $2\,Solutions$ e $0\,Solutions$ f $1\,Solution$

3 a $x=-6, x=1$ b $x=2, x=4$ c $x=4$
d $x=7, x=-2$ e $x=9, x=-1$ f $x=6$
g $x=6, x=5$ h $x=-\frac{2}{5}$ i $x=\frac{1}{2}$

4 a $x=-\frac{3}{2}, x=1$ b $x=\frac{1}{2}, x=4$ c $x=-\frac{3}{2}, x=-1$
d $x=\frac{1}{2}, x=\frac{3}{2}$ e $x=\frac{1}{4}, x=\frac{1}{3}$ f $x=\frac{1}{3}, x=\frac{2}{3}$
g $x=-\frac{5}{3}, x=2$ h $x=-\frac{1}{2}, x=\frac{1}{6}$ i $x=\frac{1}{5}, x=2$

5 a $x=\frac{-1\pm\sqrt{5}}{2}$ b $x=\frac{-5\pm\sqrt{13}}{2}$ c $x=\frac{-3\pm\sqrt{17}}{2}$
d $x=-\frac{3\pm\sqrt{33}}{2}$ e $x=\frac{3\pm\sqrt{21}}{2}$ f $x=\frac{-1\pm\sqrt{41}}{4}$
g $x=-\frac{5\pm\sqrt{57}}{4}$ h $x=\frac{9\pm\sqrt{21}}{6}$ i $x=\frac{11\pm\sqrt{13}}{6}$

6 a $x=\frac{-2\pm\sqrt{10}}{2}$ b $x=\frac{3\pm2\sqrt{3}}{3}$ c $x=\frac{5\pm\sqrt{13}}{4}$
d $x=\frac{1\pm\sqrt{7}}{2}$ e $x=\frac{-1\pm2\sqrt{2}}{2}$ f $x=\frac{3\pm2\sqrt{6}}{3}$
g $x=\frac{3\pm\sqrt{17}}{8}$ h $x=\frac{1\pm\sqrt{19}}{6}$ i $x=\frac{-1\pm\sqrt{21}}{5}$

7 a $x = \frac{5 \pm \sqrt{73}}{4}$ b $x = \frac{5 \pm \sqrt{61}}{6}$ c $x = -\frac{1 \pm \sqrt{3}}{2}$

d $x = \frac{1 \pm \sqrt{13}}{2}$ e $x = \frac{-2 \pm 3\sqrt{2}}{2}$ f $x = 2 \pm \sqrt{3}$

8 a $x = -3 \pm \sqrt{19}$ b $x = \frac{-4 \pm 2\sqrt{2}}{3}$ c $x = \frac{1 \pm \sqrt{37}}{6}$

d $x = \frac{1 \pm \sqrt{97}}{12}$ e $x = -\frac{1 \pm \sqrt{46}}{3}$ f $x = \frac{-2 \pm \sqrt{19}}{3}$

g $x = \frac{1 \pm \sqrt{11}}{5}$ h $x = \frac{-2 \pm \sqrt{13}}{3}$ i $x = \frac{5 \pm \sqrt{19}}{6}$

Challenge Yourself

1 a $x = \frac{k \pm \sqrt{k^2 - 32}}{2}$ b $x = \frac{-5 \pm \sqrt{25 + 28k}}{2k}$ c $x = \frac{-2 \pm \sqrt{4 + 3k}}{3}$

d $x = \frac{k}{2}, x = -2k$

2 $c = 7$ 3 $b = \pm 6$ 4 $k = \pm 12$

5 $k = 1, k = -3$

Did it help?

I really hope this book helped you with quadratics.

The only way I'll know is if you leave a quick review.

Could you leave a quick review?

Doesn't have to be long.

Positive.

Negative.

What you liked.

What you didn't.

That way I know what to improve.

Thanks heaps.

Made in United States
North Haven, CT
05 November 2026

21768324R10120